3D Fibre Reinforced Polymer Composites

Elsevier Science Internet Homepage - http://www.elsevier.com
Consult the Elsevier homepage for full catalogue information on all books, journals and electronic products and services.

Elsevier Titles of Related Interest

VALERY V. VASILEV & EVGENY V. MOROZOV
Mechanics and Analysis of Composite Materials
ISBN: 0 08 042702 2

JANG-KYO KIM & YIU WING MAI
Engineered Interfaces in Fiber Reinforced Composites
ISBN: 0 08 042695 6

J.G. WILLIAMS & A. PAVAN
Fracture of Polymers, Composites and Adhesives
ISBN: 0 08 043710 9

D.R. MOORE, A. PAVAN & J.G. WILLIAMS
Fracture Mechanics Testing Methods for Polymers Adhesives and Composites
ISBN: 0 08 043689 7

A. BAKER, F. ROSE & R. JONES
Advances in the Bonded Composite Repair of Metallic Aircraft
ISBN: 0 08 042699 9

Related Journals:
Composite Structures - www.elsevier.com/locate/compstruct
Composites Part A: Applied Science and Manufacturing - www.elsevier.com/locate/compositesa
Composites Part B: Engineering - www.elsevier.com/locate/compositesb
Composites Science and Technology - www.elsevier.com/locate/compscitech

Major Reference Work:
Comprehensive Composite Materials - www.elsevier.com/locate/isbn/0080429939

To contact the Publisher

Elsevier Science welcomes enquiries concerning publishing proposals: books, journal special issues, conference proceedings, etc. All formats and media can be considered. Should you have a publishing proposal you wish to discuss, please contact, without obligation, the publisher responsible for Elsevier's Composites and Ceramics programme:

Emma Hurst
Assistant Publishing Editor
Elsevier Science Ltd
The Boulevard, Langford Lane   Phone: +44 1865 843629
Kidlington, Oxford   Fax: +44 1865 843931
OX5 1GB, UK   E.mail: e.hurst@elsevier.com

General enquiries, including placing orders, should be directed to Elsevier's Regional Sales Offices – please access the Elsevier homepage for full contact details (homepage details at the top of this page).

 to search for more Elsevier books, visit the Books Butler at
http://www.elsevier.com/homepage/booksbutler/

3D Fibre Reinforced Polymer Composites

Liyong Tong
School of Aerospace, Mechanical and Mechatronic Engineering,
University of Sydney, Sydney, Australia

Adrian P. Mouritz
Department of Aerospace Engineering,
Royal Melbourne Institute of Technology, Melbourne, Australia

Michael K. Bannister
Cooperative Research Centre for Advanced Composite Structures Ltd
&
Department of Aerospace Engineering,
Royal Melbourne Institute of Technology, Melbourne, Australia

2002
ELSEVIER

AMSTERDAM – BOSTON – LONDON – NEW YORK – OXFORD – PARIS
SAN DIEGO – SAN FRANCISCO – SINGAPORE – SYDNEY – TOKYO

ELSEVIER SCIENCE Ltd
The Boulevard, Langford Lane
Kidlington, Oxford OX5 1GB, UK

First edition 2002

Library of Congress Cataloging in Publication Data
A catalog record from the Library of Congress has been applied for.

British Library Cataloguing in Publication Data
A catalogue record from the British Library has been applied for.

ISBN: 0-08-043938-1

⊗ The paper used in this publication meets the requirements of ANSI/NISO Z39.48-1992 (Permanence of Paper).
Printed in the Netherlands.

To my wife Hua and my children Richard and Victoria   L. Tong

To my wife Jenny and my children Lauren and Christian   A.P. Mouritz

To my wife Ruth and my children Lachlan and Emma   M.K. Bannister

Preface

Fibre reinforced polymer (FRP) composites are used in almost every type of advanced engineering structure, with their usage ranging from aircraft, helicopters and spacecraft through to boats, ships and offshore platforms and to automobiles, sports goods, chemical processing equipment and civil infrastructure such as bridges and buildings. The usage of FRP composites continues to grow at an impressive rate as these materials are used more in their existing markets and become established in relatively new markets such as biomedical devices and civil structures. A key factor driving the increased applications of composites over recent years is the development of new advanced forms of FRP materials. This includes developments in high performance resin systems and new styles of reinforcement, such as carbon nanotubes and nanoparticles. A major driving force has been the development of advanced FRP composites reinforced with a three-dimensional (3D) fibre structure. 3D composites were originally developed in the early 1970s, but it has only been in the last 10-15 years that major strides have been made to develop these materials to a commercial level where they can be used in both traditional and emerging markets.

The purpose of this book is to provide an up-to-date account of the fabrication, mechanical properties, delamination resistance, impact damage tolerance and applications of 3D FRP composites. The book will focus on 3D composites made using the textile technologies of weaving, braiding, knitting and stitching as well as by z-pinning. This book is intended for undergraduate and postgraduate students studying composite materials and also for the researchers, manufacturers and end-users of composites.

Chapter 1 provides a general introduction to the field of advanced 3D composites. The chapter begins with a description of the key economic and technology factors that are providing the impetus for the development of 3D composites. These factors include lower manufacturing costs, improved material quality, high through-thickness properties, superior delamination resistance, and better impact damage resistance and post-impact mechanical properties compared to conventional laminated composites. The current and potential applications of 3D composites are then outlined in Chapter 1, including a description of the critical issues facing their future usage.

Chapter 2 gives a description of the various weaving, braiding, knitting and stitching processes used to manufacture 3D fabrics that are the preforms to 3D composites. The processes that are described range from traditional textile techniques that have been used for hundreds of years up to the most recent textile processes that are still under development. Included in the chapter is an examination of the affect the processing parameters of the textile techniques have on the quality and fibre architecture of 3D composites.

The methods and tooling used to consolidate 3D fabric preforms into FRP composites are described in Chapter 3. The liquid moulding methods used for consolidation include resin transfer moulding, resin film infusion and SCRIMP. The benefits and limitations of the different consolidation processes are compared for producing 3D composites. Chapter 3 also gives an overview of the different types of processing defects (eg. voids, dry spots, distorted binder yarns) that can occur in 3D composites using liquid moulding methods.

A review of micro-mechanical models that are used or have a potential to be used to theoretically analyse the mechanical properties of 3D textile composites is presented in Chapter 4. Models for determining the in-plane elastic modulus of 3D composites are described, including the Eshlby, Mori-Tanaka, orientation averaging, binary and unit cell methods. Models for predicting the failure strength are also described, such as the unit cell, binary and curved beam methods. The accuracy and limitations of models for determining the in-plane properties of 3D composites are assessed, and the need for more reliable models is discussed.

The performance of 3D composites made by weaving, braiding, knitting, stitching and z-pinning are described in Chapters 5 to 9, respectively. The in-plane mechanical properties and failure mechanisms of 3D composites under tension, compression, bending and fatigue loads are examined. Improvements to the interlaminar fracture toughness, impact resistance and damage tolerance of 3D composites are also described in detail. In these chapters the gaps in our understanding of the mechanical performance and through-thickness properties of 3D composites are identified for future research.

We thank our colleagues with whom we have researched and developed 3D composites over the last ten years, in particular to Professor I. Herszberg, Professor G.P. Steven, Dr P. Tan, Dr K.H. Leong, Dr P.J. Callus, Dr P. Falzon, Mr K. Houghton, Dr L.K. Jain and Dr B.N. Cox. We are thankful to many colleagues, in particular to Professors T.-W. Chou, O.O. Ochoa, and P. Smith, for their kind encouragement in the initiation of this project. We are indebted to the University of Sydney, the Royal Melbourne Institute of Technology and the Cooperative Research Centre for Advanced Composite Structures Ltd. for allowing the use of the facilities we required in the preparation of this book. LT and APM are grateful for funding support of the Australian Research Council (Grant No. C00107070, DP0211709), Boeing Company, and Boeing (Hawker de Havilland) as well as the Cooperative Research Centre for Advanced Composite Structures Ltd. We are also thankful to the many organisations that kindly granted permission to use their photographs, figures and diagrams in the book.

L. Tong
School of Aerospace, Mechanical & Mechatronic Engineering
University of Sydney

A.P. Mouritz
Department of Aerospace Engineering
Royal Melbourne Institute of Technology

M.K. Bannister
Cooperative Research Centre for Advanced Composite Structures Ltd
&
Department of Aerospace Engineering
Royal Melbourne Institute of Technology

Table of Contents

Chapter 1

Introduction

1.1 BACKGROUND

Fibre reinforced polymer (FRP) composites have emerged from being exotic materials used only in niche applications following the Second World War, to common engineering materials used in a diverse range of applications. Composites are now used in aircraft, helicopters, space-craft, satellites, ships, submarines, automobiles, chemical processing equipment, sporting goods and civil infrastructure, and there is the potential for common use in medical prothesis and microelectronic devices. Composites have emerged as important materials because of their light-weight, high specific stiffness, high specific strength, excellent fatigue resistance and outstanding corrosion resistance compared to most common metallic alloys, such as steel and aluminium alloys. Other advantages of composites include the ability to fabricate directional mechanical properties, low thermal expansion properties and high dimensional stability. It is the combination of outstanding physical, thermal and mechanical properties that makes composites attractive to use in place of metals in many applications, particularly when weight-saving is critical.

FRP composites can be simply described as multi-constituent materials that consist of reinforcing fibres embedded in a rigid polymer matrix. The fibres used in FRP materials can be in the form of small particles, whiskers or continuous filaments. Most composites used in engineering applications contain fibres made of glass, carbon or aramid. Occasionally composites are reinforced with other fibre types, such as boron, Spectra® or thermoplastics. A diverse range of polymers can be used as the matrix to FRP composites, and these are generally classified as thermoset (eg. epoxy, polyester) or thermoplastic (eg. polyether-ether-ketone, polyamide) resins.

In almost all engineering applications requiring high stiffness, strength and fatigue resistance, composites are reinforced with continuous fibres rather than small particles or whiskers. Continuous fibre composites are characterised by a two-dimensional (2D) laminated structure in which the fibres are aligned along the plane (x- & y-directions) of the material, as shown in Figure 1.1. A distinguishing feature of 2D laminates is that no fibres are aligned in the through-thickness (or z-) direction. The lack of through-thickness reinforcing fibres can be a disadvantage in terms of cost, ease of processing, mechanical performance and impact damage resistance.

A serious disadvantage is that the current manufacturing processes for composite components can be expensive. Conventional processing techniques used to fabricate composites, such as wet hand lay-up, autoclave and resin transfer moulding, require a high amount of skilled labour to cut, stack and consolidate the laminate plies into a preformed component. In the production of some aircraft structures up to 60 plies of carbon fabric or carbon/epoxy prepreg tape must be individually stacked and aligned by hand. Similarly, the hulls of some naval ships are made using up to 100 plies of woven glass fabric that must be stacked and consolidated by hand. The lack of a z-direction binder means the plies must be individually stacked and that adds considerably to the fabrication time. Furthermore, the lack of through-thickness fibres means that the plies can slip during lay-up, and this can misalign the fibre orientations in the composite component. These problems can be alleviated to some extent by semi-automated processes that reduce the amount of labour, although the equipment is very expensive and is often only suitable for fabricating certain types of structures, such as flat and slightly curved panels. A further problem with fabricating composites is that production rates are often low because of the slow curing of the resin matrix, even at elevated temperature.

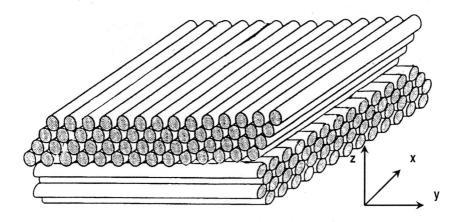

Figure 1.1 Schematic of the fibre structure to a 2D laminate

Fabricating composites into components with a complex shape increases the cost even further because some fabrics and many prepreg tapes have poor drape. These materials are not easily moulded into complex shapes, and as a result some composite components need to be assembled from a large number of separate parts that must be joined by co-curing, adhesive bonding or mechanical fastening. This is a major problem for the aircraft industry, where composite structures such as wing sections must be made from a large number of smaller laminated parts such as skin panels, stiffeners and stringers. These fabrication problems have impeded the wider use of composites in some aircraft structures because it is significantly more expensive than using aircraft-grade aluminium alloys.

As well as high cost, another major disadvantage of 2D laminates is their low through-thickness mechanical properties because of the lack of z-direction fibres. The two-dimensional arrangement of fibres provides very little stiffness and strength in the through-thickness direction because these properties are determined by the low mechanical properties of the resin and fibre-to-resin interface. A comparison of the in-plane and through-thickness strengths of 2D laminates is shown in Figure 1.2. It is seen that the through-thickness properties are often less than 10% of the in-plane properties, and for this reason 2D laminates can not be used in structures supporting high through-thickness or interlaminar shear loads.

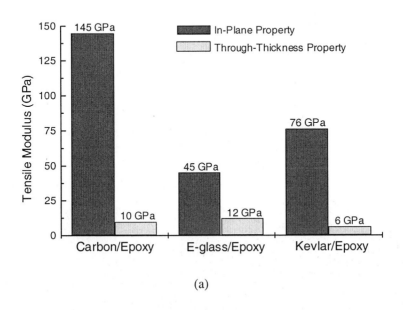

(a)

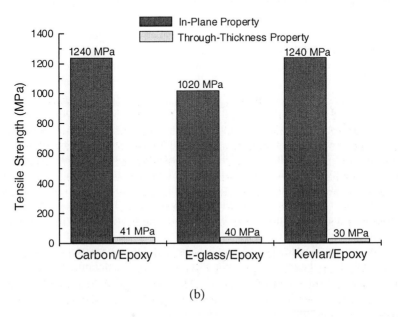

(b)

Figure 1.2 Comparison of in-plane and through-thickness mechanical properties of some engineering composites.

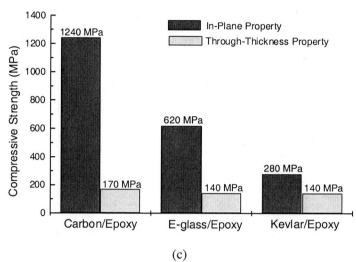

(c)

Figure 1.2 (continued) Comparison of in-plane and through-thickness mechanical properties of some engineering composites.

A further problem with 2D laminates is their poor impact damage resistance and low post-impact mechanical properties. Laminates are prone to delamination damage when impacted by low speed projectiles because of the poor through-thickness strength. This is a major concern with composite aircraft structures where tools dropped during maintenance, bird strikes, hail impacts and stone impacts can cause damage. Similarly, the composite hulls to yachts, boats and ships can be damaged by impact with debris floating in the water or striking a wharf while moored in heavy seas. This damage can be difficult to detect, particularly when below the waterline, and can affect water-tightness and structural integrity of the hull. Impact damage can seriously degrade the in-plane mechanical properties under tension, compression, bending and fatigue loads. For example, the effect of impact loading on the tension and compression strengths of an aerospace grade carbon/epoxy laminate is shown in Figure 1.3. The strength drops rapidly with increasing impact energy, and even a lightweight impact of several joules can cause a large loss in strength. The low post-impact mechanical properties of 2D laminates is a major disadvantage, particularly when used in thin load-bearing structures such as aircraft fuselage and wing panels where the mechanical properties can be severely degraded by a small amount of damage. To combat the problem of delamination damage, composite parts are often over-designed with extra thickness. However, this increases the cost, weight and volume of the composite and in some cases may provide only moderate improvements to impact damage resistance.

Various materials have been developed to improve the delamination resistance and post-impact mechanical properties of 2D laminates. The main impact toughening methods are chemical and rubber toughening of resins, chemical and plasma treatment of fibres, and interleaving using tough thermoplastic film. These methods are effective in improving damage resistance against low energy impacts, although each has a number of drawbacks that has limited their use in large composite structures. The greatest drawback with most of the methods is the high cost. For example, toughened resins are usually at least 5-10 times more expensive than standard resins. In the case of interleaving it is necessary to cut and individually arrange the interleaf films between the fabric plies before infusing the resin. The inclusion of the interleaf films is a laborious and slow process that can add considerably to the manufacturing cost of a composite. Another problem is the difficultly in manufacturing high-quality laminates using some of the methods. For example, in the manufacture of composites with rubber toughened resin, the fine rubber particles can become trapped within the fabric reinforcement and may not be evenly distributed through the material.

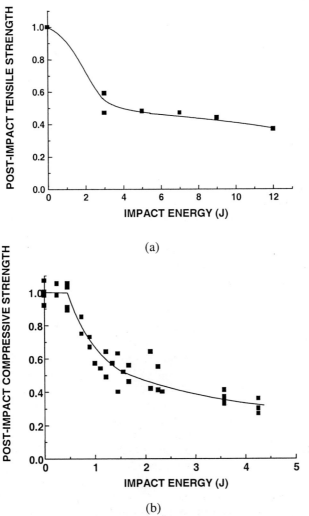

(a)

(b)

Figure 1.3 Effect of impact energy on the (a) residual tensile strength and (b) residual compressive strength of 2D carbon/epoxy laminate. The post-impact strength values are normalised to the strength of the laminate without impact damage (The tensile and compressive strength data are from Dorey (1989) and Caprino (1984), respectively).

1.2 INTRODUCTION TO 3D FRP COMPOSITES

Since the late-1960s, various types of composite materials with three-dimensional (3D) fibre structures (incorporating z-direction fibres) have been developed to overcome the shortcomings of 2D laminates. That is, the development of 3D composites has been driven by the needs to reduce fabrication cost, increase through-thickness mechanical properties and improve impact damage tolerance. The development of 3D composites has been undertaken largely by the aerospace industry due to increasing demands on FRP materials in load-bearing structures to aircraft, helicopters and space-craft. The marine, construction and automotive industries have supported the developments. 3D composites are made using the textile processing techniques of weaving, knitting, braiding and stitching. 3D composites are also made using a novel process known as z-pinning.

Braiding was the first textile process used to manufacture 3D fibre preforms for composite. Braiding was used in the late 1960s to produce 3D carbon-carbon composites to replace high temperature metallic alloys in rocket motor components in order to reduce the weight by 30-50% (Stover et al., 1971). An example of a modern rocket nozzle fabricated by 3D braiding is shown in Figure 1.4. At the time only a few motor components were made, although it did demonstrate the capability of the braiding process to produce intricately shaped components from advanced 3D composites. Shortly afterwards, weaving was used for the first time to produce 3D carbon-carbon composites for brake components to jet aircraft (Mullen and Roy, 1972). 3D woven composites were made to replace high-temperature metal alloys in aircraft brakes to improve durability and reduce heat distortion.

Figure 1.4 3D braided preform for a rocket nozzle (Courtesy of the Atlantic Research Corporation)

It is worth noting that these early 3D composites were made of carbon-carbon materials and not fibre reinforced polymers. The need for 3D FRP composites was not fully appreciated in the 1960s, and it was not until the mid-1980s that development commenced on these materials. From 1985 to 1997 a NASA-lead study known as the 'Advanced Composite Technology Program' (ACTP), that included participants from aircraft companies, composite suppliers and the textiles industry, was instrumental in the research and development of 3D FRP composites (Dow and Dexter, 1997). The program examined the potential of the textile processes of weaving, braiding, knitting and stitching to produce advanced 3D composites for aircraft components. Developmental work from the ACTP, combined with studies performed by other research institutions, has produced an impressive variety of components and structures made using 3D composites, and some of these are described below. However, due to the commercial sensitivity of some components only those reported in the open literature will be described.

1.2.1 Applications of 3D Woven Composites

Weaving is a process that has been used for over 50 years to produce single-layer, broad-cloth fabric for use as fibre reinforcement to composites. It is only relatively recently, however, that weaving techniques have been modified to produce 3D woven materials that contain through-thickness fibres binding together the in-plane fabrics. A variety of 3D woven composites have been manufactured using modified weaving looms with different amounts of x-, y- and z-direction fibres so that the properties can be tailored to a specific application. The great flexibility of the 3D weaving process means that a wide variety of composite components have been developed for aerospace, marine, civil infrastructure and medical applications (Mouritz et al., 1999). However, only a few 3D woven components are currently used; most of the components have been manufactured as demonstration items to showcase the potential applications of 3D woven composites. A list of applications for 3D woven composites is given in Table 1.1 and some woven preform structures are shown in Figure 1.5. It is seen that a range of intricate shapes can be integrally woven for possible applications as flanges, turbine rotors, beams and cylinders. In the production of these demonstration items it has been proven in many cases that it is faster and cheaper to manufacture 3D woven components than 2D laminates, particularly for complex shapes. Furthermore, 3D woven components have superior delamination resistance and impact damage tolerance.

Table 1.1 Demonstrator components made with 3D woven composite

Turbine engine thrust reversers, rotors, rotor blades, insulation, structural reinforcement and heat exchangers
Nose cones and nozzles for rockets
Engine mounts
T-section elements for aircraft fuselage frame structures
Rib, cross-blade and multi-blade stiffened aircraft panels
T- and X-shape elements for filling the gap at the base of stiffeners when manufacturing stiffened panels
Leading edges and connectors to aircraft wings
I-beams for civil infrastructure
Manhole covers

(a)

(b)

Figure 1.5 Examples of 3D woven preforms. (a) Cylinder and flange, (b) egg crate structures and (c) turbine rotors woven by the Techniweave Inc. (Photographs courtesy of the Techniweave Inc.).

(c)

Figure 1.5 (continued) Examples of 3D woven preforms. (a) Cylinder and flange, (b) egg crate structures and (c) turbine rotors woven by the Techniweave Inc. (Photographs courtesy of the Techniweave Inc.).

While a variety of components have been made to demonstrate the versatility and capabilities of 3D weaving, the reported applications for the material are few. One application is the use of 3D woven composite in H-shaped connectors on the Beech starship (Wong, 1992). The woven connectors are used for joining honeycomb wing panels together. 3D composite is used to reduce the cost of manufacturing the wing as well as to improve stress transfer and reduce peeling stresses at the joint.

3D woven composite is being used in the construction of stiffeners for the air inlet duct panels to the Joint Strike Fighter (JSF) being produced by Lockheed Martin. The use of 3D woven stiffeners eliminates 95% of the fasteners through the duct, thereby improving aerodynamic and signature performance, eliminating fuel leak paths, and simplifying manufacturing assembly compared with conventional 2D laminate or aluminium alloy. It is estimated the ducts can be produced in half the time and at two-thirds the cost of current inlet ducts, and save 36 kg in weight and at least US$200,000 for each duct.

3D woven composite is also being used in rocket nose cones to provide high temperature properties, delamination and erosion resistance compared with traditional 2D laminates. It is estimated that the 3D woven nose cones are produced at about 15% of the cost of conventional cones, resulting in significant cost saving. 3D woven sandwich composites are being used in prototype Scramjet engines capable of speeds up to Mach 8 (~2600 m/s) (Kandero, 2001). The 3D material is a ceramic-based composite consisting of 3D woven carbon fibres in a silicon carbide matrix. The 3D composite is used in the combustion chamber to the Scramjet engine. A key benefit of using 3D woven composite is the ability to manufacture the chamber as a single piece by 3D weaving, and this reduces connection issues and leakage problems associated with conventional fabrication methods.

Apart from these aerospace applications, the only other uses of 3D woven composite is the very occasional use in the repair of damaged boat hulls, I-beams in the roof of a ski chair-lift building in Germany (Müller et al., 1994), manhole covers, sporting goods such as shin guards and helmets, and ballistic protection for police and military personnel (Mouritz et al., 1999). 3D woven composite is not currently used as a biomedical material, although its potential use in leg prosthesis has been explored (Limmer et al., 1996).

1.2.2 Applications of 3D Braided Composites

The braiding process is familiar to many fields of engineering as standard 2D braided carbon and glass fabric have been used for many years in a variety of high technology items, such as golf clubs, aircraft propellers and yacht masts (Popper, 1991). 3D braided preform has a number of important advantages over many types of 2D fabric preforms and prepreg tapes, including high levels of conformability, drapability, torsional stability and structural integrity. Furthermore, 3D braiding processes are capable of forming intricately-shaped preforms and the process can be varied during operation to produce changes in the cross-sectional shape as well as to produce tapers, holes, bends and bifurcations in the final preform.

Potential aerospace applications for 3D braided composites are listed in Table 1.2, and include airframe spars, F-section fuselage frames, fuselage barrels, tail shafts, rib stiffened panels, rocket nose cones, and rocket engine nozzles (Dexter, 1996; Brown, 1991; Mouritz et al., 1999). A variety of other components have been made of 3D braided composite as demonstration items, including I-beams (Yau et al., 1986; Brown, 1991; Chiu et al., 1994; Fukuta, 1995; Wulfhorst et al., 1995), bifurcated beams (Popper and McConnell, 1987), connecting rods (Yau et al., 1986), and C-, J- and T-section panels (Ko, 1984; Crane and Camponesch, 1986; Macander et al., 1986; Gause and Alper, 1987; Popper and McConnell, 1987; Malkan and Ko, 1989; Brookstein, 1990; Brookstein, 1991; Fedro and Willden, 1991; Gong and Sankar, 1991; Brookstein, 1993; Dexter, 1996).

Table 1.2 Demonstrator components made with 3D braided composite.

Airframe spars, fuselage frames and barrels
Tail shafts
Rib-stiffened, C-, T- and J-section panels
Rocket nose cones and engine nozzles
Beams and trusses
Connecting rods
Ship propeller blades
Biomedical devices

In the non-aerospace field, 3D braided composite has been used in propeller blades for a naval landing craft (Maclander et al., 1986; Maclander, 1992). There is also potential application for 3D braided composite on ships, such as in propulsion shafts and propellers (Mouritz et al., 2001). 3D braided composite has been used in truss section decking for light-weight military bridges capable of carrying tanks and tank carriers (Loud, 1999). Other potential applications include military landing pads, causeways, mass transport and highway bridge structures when bonded to pre-stressed concrete. 3D braided composite also has potential uses in the bodies, chassis and drive shafts of automobiles because they are about 50% lighter than the same components made of steel but with similar damage tolerance and crashworthiness properties (Brandt and Drechsler, 1995). 3D braided composite has also been manufactured into a number of biomedical devices (Ko et al., 1988).

1.2.3 3D Knitted Composites

3D knitted composite has a number of important advantages over conventional 2D laminate, particularly very high drape properties and superior impact damage resistance. Despite these advantages, there are some drawbacks with 3D knitted material that has limited its application. A number of aircraft structures have been made of 3D knitted composite to demonstrate the potential of these materials, such as in wing stringers (Clayton et al., 1997), wing panels (Dexter, 1996), jet engine vanes (Gibbon, 1994; Sheffer & Dias, 1998), T-shape connectors (King et al., 1996) and I-beams (Sheffer & Dias, 1998). This composite is under investigation for the manufacture of the rear pressure bulkhead to the new Airbus A380 aircraft (Hinrichsen, 2000). The potential use of 3D knitted composite in non-aerospace components includes bumper bars, floor panels and door members for automobiles (Hamilton and Schinske, 1990), rudder tip fairings, medical prothesis (Mouritz et al., 1999), and bicycle helmets (Verpoest et al., 1997).

1.2.4 3D Stitched Composites

The stitching of laminates in the through-thickness direction with a high strength thread has proven a simple, low-cost method for producing 3D composites. Stitching basically involves inserting a fibre thread (usually made of carbon, glass or Kevlar) through a stack of prepreg or fabric plies using an industrial grade sewing machine. The amount of through-thickness reinforcement in stitched composites is normally between 1 to 5%, which is a similar amount of reinforcement in 3D woven, braided and knitted composites.

Stitching is used to reinforce composites in the z-direction to provide better delamination resistance and impact damage tolerance than conventional 2D laminates. Stitching can also be used to construct complex three-dimensional shapes by stitching a number of separate composite components together. This eliminates the need for mechanical fasteners, such as rivets, screws and bolts, and thereby reduces the weight and possibly the production cost of the component. If required, stitches can be placed only in areas that would benefit from through-thickness reinforcement, such as along the edge of a composite component, around holes, cut-outs or in a joint.

A variety of 3D composite structures have been manufactured using stitching, and the more important stitched structures are lap joints, stiffened panels, and aircraft wing-

mechanically or electronically operated and may allow individual yarns to be selectively controlled (jacquard loom) or control a set of yarns simultaneously (loom with shafts, as shown in Figure 2.1). The crucial point is that the lifting mechanism selects and lifts the required yarns and creates a space (the shed) into which the weft yarns are inserted at right angles to the warp (the 90° direction). The sequence in which the warp yarns are lifted controls the interlinking of the warp and weft yarns and thus the pattern that is created in the fabric (see Figure 2.2). It is this pattern that influences many of the fabric properties, such as mechanical performance, drapability, and fibre volume fraction. Therefore to manufacture a suitable 2D or 3D preform an understanding of how the required fibre architecture can be produced through the design of the correct lifting pattern is crucial in the use of this manufacturing process.

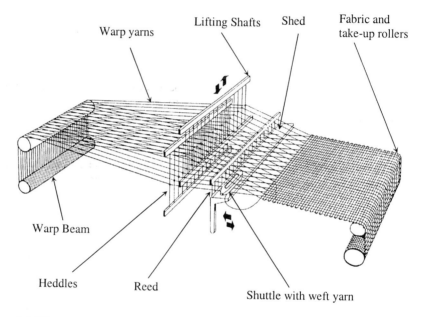

Figure 2.1 Illustration of conventional weaving process

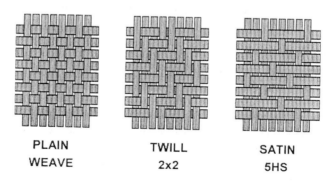

Figure 2.2 Typical 2D weave patterns

The insertion of the weft yarns can be done using a number of methods. One of the oldest techniques consists of transferring a small package of yarn in a holder (shuttle) through the shed, the yarn being drawn out of the shuttle and laid across the warp yarns as the shuttle moves. This is a relatively slow technique but has the advantage of creating a closed edge to the fabric, as it is a single continuous yarn that is forming the fabric weft. More recent, high-speed techniques involve laying down separate weft yarns across the fabric width. These weft yarns are drawn through the shed mechanically with a long slender arm (rapier) or pushed across with high-pressure bursts of air or water. These processes are faster than shuttle looms, reaching speeds of approximately 600 insertions/minute, but create a loose edge of cut weft yarns that needs to be bound together so that the fabric does not fray (salvage).

The final mechanism involved in the weaving process is a comb-like device (reed) that is used to correctly space the warp yarns across the width of the fabric and to compact the fabric after the weft yarns have been inserted. Generally a series of positively driven rollers are used to pull the fabric out of the loom as it is being produced and to provide a level of fabric tension during the weaving process. It should be noted that the resultant fabric consists only of 0° and 90° yarns, conventional weaving is incapable of producing fabrics with yarns at any other angles and this is one of the main disadvantages of weaving over other textile processes.

Current, commercial looms generally produce fabric of widths between 1.8 – 2.5 metres at production rates of metres/minute. The standard weaving process is therefore ideally suited to the cost-effective production of large volumes of material. However, using essentially the same equipment, the process described above can also be used to produce more complex, multilayer fabrics that have yarns in the thickness direction linking the layers together.

2.2.2 Multilayer or 3D Weaving

The first major difference between conventional weaving and multilayer weaving is the requirement to have multiple layers of warp yarns. The greater the number of layers required (and thus the thickness of the preform) or the wider the fabric produced, means a larger number of individual warp yarns that have to be fed into the loom and controlled during the lifting sequence. Therefore the source of the warp yarn for multilayer weaving is generally from large creels in which each warp yarn comes from its own individual yarn package. Multiple warp beam systems have also been used although this is not as common. This requirement for a large number of warp ends raises the first disadvantage of weaving, namely that the cost of obtaining (generally) thousands of yarns packages and the time required to set up the large number of warp ends within the loom can be extremely expensive. This non-recurring cost becomes less significant as the length of the fabric being woven increases but having to weave large volumes of the same material restricts the flexibility of the process. Most multilayer weaving is therefore currently used to produce relatively narrow width products, where the number of warp ends is relatively small, or high value products where the cost of the preform production is acceptable.

As most 3D composites are produced from high performance yarns (carbon, glass, ceramic, etc) standard textile tensioning rollers are unsuitable and tension control on the individual yarns during the weaving is critical in obtaining a consistent preform quality. This is generally accomplished through spring-loaded or frictional tension devices on the creel or through hanging small weights on the yarns before entering the lifting device. Figure 2.3 illustrates the use of multiple warp beams and hanging weights in multilayer weaving. The lifting mechanisms are the same as used in conventional weaving although the heddle eyes through which the yarn passes tend to be smoothed and rounded to minimise friction with the more brittle high performance fibres. Jacquard lifting mechanisms tend to be used more frequently as their greater control over individual warp yarns offers more flexibility in the weave patterns produced. The weft insertion is accomplished with standard technology (generally a rapier mechanism) inserting individual wefts between the selected warp layers. Variations in the lifting and weft insertion mechanisms to allow multiple sheds to be formed and thus multiple simultaneous weft insertions have also been developed and would allow a faster preform production rate. This type of technology is often regarded as the true 3D weaving.

Figure 2.3 Multilayer weaving loom (courtesy of the Cooperative Research Centre for Advanced Composite Structures, Ltd)

It is through the design of the lifting pattern that the three-dimensional nature of the weave architecture is produced in multilayer weaving. Commonly the bulk of the warp and weft yarns are designed to lay straight within the preform and thus maximise the mechanical performance. In order to bind the preform together, selected warp yarns, coming from a separate beam if warp beams are used, are lifted and dropped so that their path travels in the thickness direction thus binding the layers together (Figure 2.4).

Such a multilayer weaving loom is described by Yamamoto et al (1995). Examples of such weave architectures that are currently capable of being manufactured using multilayer weaving are illustrated in Figure 2.5. It should be noted that the illustrations in Figure 2.5 show idealised architectures and often these can be very different from the resultant preform architecture (Bannister et al 1998). Tension within and friction between the yarns, together with the initial weave parameters (yarn size and twist, yarn spacing, number of layers, etc) can all affect the final architecture and thus the composite performance. As with conventional weaving, multilayer weaving is only capable of producing fabrics with 0° and 90° in-plane yarns, although the binder yarns can be oriented at an angle. This tends to limit the use of these preforms as their shear and torsional properties can be relatively low. Various 3D weaving techniques can produce preforms with yarns at other angles although this requires the use of highly specialised equipment, which will be discussed later.

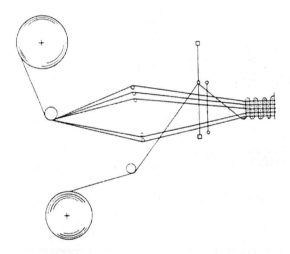

Figure 2.4 Illustration of multilayer weaving

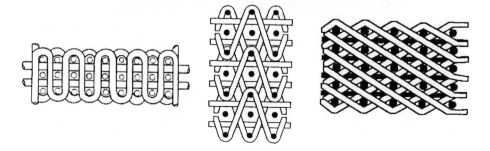

Figure 2.5 Typical multilayer yarn architectures

Flat, multilayer fabrics are not the only structures that can be woven on standard looms. By correctly programming the sequence in which the warp yarns are lifted it is possible to weave a fabric with slits that can be opened out to form a complex three-dimensional structure. This concept is illustrated in Figure 2.6, which demonstrates how I-beams and box structures can be formed from, initially, flat fabric. An example of such an integrally woven I-beam is shown in Figure 2.7 and these types of components have already been used in the civil engineering field (Müller et al., 1994). A reasonable range of shaped products can be formed in such a way however more advanced forms of 3D weaving are capable of producing more complex preforms.

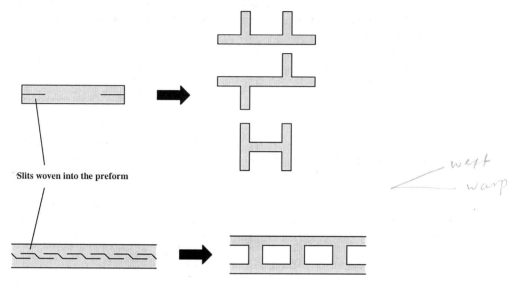

Slits woven into the preform

Figure 2.6 Production of shaped components from flat multilayer preforms

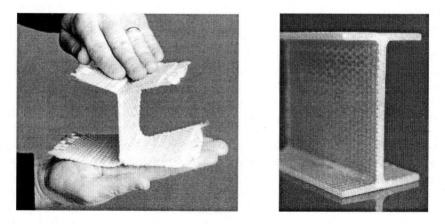

Figure 2.7 Formation of composite I-beam from a flat multilayer preform (courtesy of the Cooperative Research Centre for Advanced Composite Structures, Ltd)

In spite of some limitations in preform design with the multilayer weaving process, its greatest advantage is that it can be performed upon conventional weaving looms and does not require significant costs to develop specialised machinery. It appears suited primarily to the manufacture of large volumes of flat or simply shaped preforms with a basic 0° and 90° yarn architecture.

2.2.3 3D Orthogonal Non-Wovens

There is still some argument as to what constitutes the distinction between multilayer (or 3D weaving) and 3D orthogonal non-wovens. The traditional definition of weaving requires the yarns to be interlaced with each other, thus processes that produce preforms with the yarns in orthogonal, non-interlaced architectures are generally referred to as 3D orthogonal non-wovens (Khokar, 1996). These processes generally differ from multilayer weaving in that multiple yarns that are separate from the warp yarns (X direction) are inserted in the Y and Z directions in a highly controlled manner. The production of a 3D fibre architecture using a 3D non-woven process therefore does not solely rely upon the warp yarn lifting sequence. Confusion can sometimes occur due to the fact that 3D weaving equipment is also capable of producing orthogonal non-woven preforms through the selection of a suitable lifting sequence. It would therefore be better to define the style of preform produced rather than the equipment used in manufacture, however this is not yet the case in the majority of the literature.

Since the 1970's a wide range of processes have been developed to produce 3D orthogonal preforms. These vary from techniques utilising relatively conventional weaving mechanisms but with multiple weft insertion (Mohamed et al., 1988), to processes (Mohamed et al., 1988; Ko, 1989a) that have very little in common with the traditional weaving process. Some of the earliest work in 3D orthogonal nonwovens was pioneered in France by Aerospatiale and Brochier who licensed their separately developed technology in the USA to Hercules (Btuno et al., 1986) and Avco Speciality Materials (Rolincik, 1987; Mullen and Roy, 1972; McAllister, and Taverna, 1975) respectively. Both processes are similar in that they use an initial framework around which radial and circumferential yarns (for cylindrical preforms) or Y and Z yarns (for rectangular billets) are placed. For the Brochier process (Autoweave™) this framework consists of pre-cured reinforcements inserted into a phenolic foam mandrel whilst the Aerospatiale process uses a network of metallic rods and plates that are removed during the placement of the axial yarns (see Figure 2.8). Both processes are capable of producing shaped preforms by suitable shaping of the initial framework and can be used to construct 4D and 5D preforms, that is with architectures containing fibres laid in directions other than X, Y or Z. These two processes have been mostly used for the production of carbon/carbon composites for use as components in rocket motors and exit cones.

Significant development of machinery to manufacture 3D non-woven preforms has also been undertaken within Japan since the 1970's, particularly at the Three-D Composites Research Corporation (a subsidiary of the Mitsubishi Electric Corporation). Methods for the production of non-woven preforms have been developed by Fukuta et al. (1974) and Kimbara et al (1991), an example of which is shown in Figure 2.9. Again these processes rely upon the insertion of yarn or cured composite rods along pre-set directions, the main difference between these methods and others being the mechanisms to control that insertion.

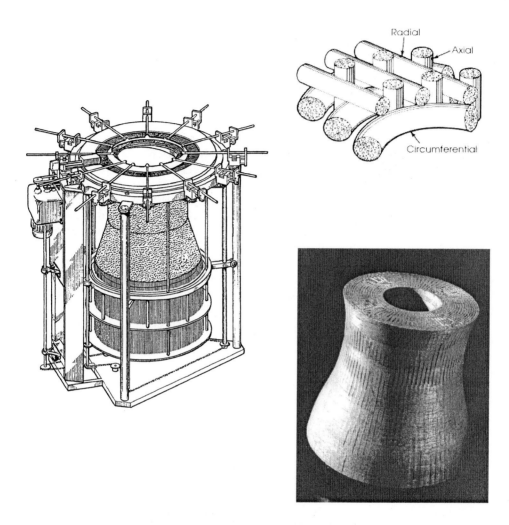

Figure 2.8 Illustration of Aerospatiale method for producing 3D orthogonal non-woven preforms and an example of a consolidated preform

Unlike multiaxial weaving, orthogonal non-woven processes are more capable of producing yarn architectures close to the idealised view, although they are generally a slower production method than those utilising more conventional weaving technology. Although the processes described here can produce a very wide variety of preforms that are generally more complex than those produced via multilayer weaving, the commercial use of these processes has been extremely limited. Most of the equipment that has been developed is highly specialised and generally not suited for large volume production, thus its commercial use has been primarily in the production of expensive carbon/carbon or ceramic composite structures.

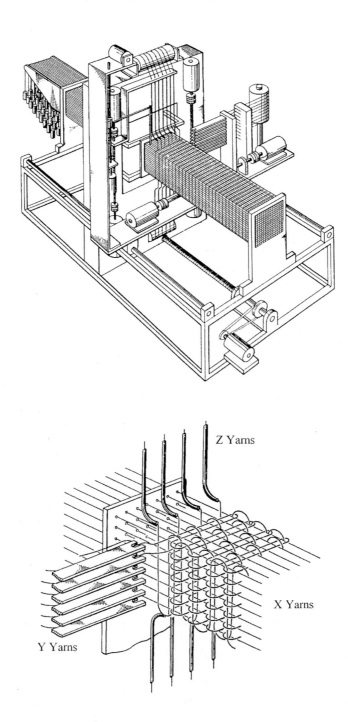

Z Yarns

X Yarns

Y Yarns

Figure 2.9 Illustration of Fukuta's et al. (1974) equipment for the manufacture of 3D non-woven preforms.

2.2.4 Multiaxial Weaving

One of the main problems facing the use of multilayer woven fabrics is the difficulty in producing a fabric that contains fibres orientated at ±45° in the plane of the preform. Standard industry looms, which are capable of producing multilayer fabric, cannot manufacture this fabric with fibres at angles other than 0° and 90°. It is possible to orient the through-thickness binder yarns at angles such as ±45° but this will not significantly affect the in-plane, off-axis properties of the composite. Although some orthogonal non-woven preforms can be produced with yarn architectures of this type, the equipment and processes used in their production are generally not suited for large volume production. This restricts the potential components that can be made using multilayer fabric as the necessity to add ±45° fabric will often negate the advantages that can be gained in using a single, integrally woven preform that contains fibres in the thickness direction. The more recent machinery developments have therefore tended to concentrate upon the formation of preforms with multiaxial yarns.

Curiskis et al (1997) have reviewed and described the techniques that are being employed to produce multiaxial preforms. Process such as Triaxial Weaving, Lappet Weaving and Split Reed Systems have been used by a number of researchers to develop equipment capable of producing multiaxial, multilayer preforms and a number of patents have been filed relating to the development of this equipment (Ruzand and Guenot, 1994; Farley, 1993; Anahara et al., 1991; Addis, 1996; Mohamed and Bilisik, 1995). Although promising results have been demonstrated, the current reported technology still appears to be in the development stage and preforms seem limited to having the ±45° yarns only towards the outer surfaces and not at other levels within the thickness of the preform (see Figure 2.10).

2.2.5 Distance Fabrics

A final subset of the weaving technologies relates to the production of a preform style known generally as Distance Fabric. This family of preforms is produced by the use of the traditional textile technique known as Velvet Weaving. In this multilayer weaving process two sets of warp yarns, spaced by a fixed distance, are woven as separate fabrics but are also interlinked by the transfer of specific warp yarns from one fabric layer to the other. These warp yarns, known as pile yarns, are woven into each face fabric thus forming a strong linkage between the two faces and creating a sandwich structure as shown in Figure 2.11. The spacing between the face fabrics can be adjusted by controlling the separation of the warp yarns in the weaving loom and the angle of the pile yarns can be varied from vertical (90°) to bias angles (e.g. ±45°) although currently these bias angles can be only produced in the warp direction. Distance Fabric material is commercially available and comes in a range of heights up to ~ 23 mm. Due to the strong linkage between the face fabrics it is highly suited for the production of peel-resistant and delamination resistant sandwich structures (Bannister et al., 1999).

2.3 BRAIDING

The braiding process is familiar to many fields of engineering as standard two-dimensionally braided carbon and glass fabric has been used for a number of years in a variety of high technology items, such as: golf clubs, aircraft propellers, yacht masts and light weight bridge structures (Popper, 1991). Thick, multilayered preforms can be manufactured through traditional 2D braiding, but the processes of 2D and 3D braiding and the variety of possible preforms that can be manufactured using these techniques are generally very different.

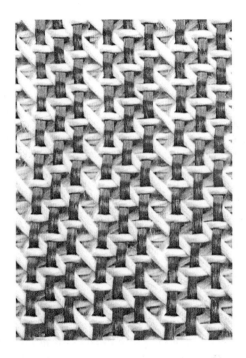

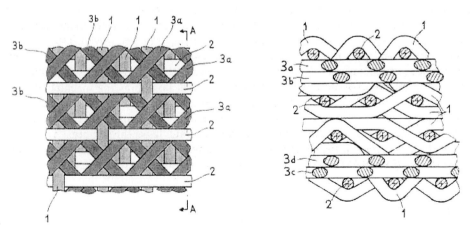

Figure 2.10 Example of multilayer woven fabric containing 0°, 90° and ±45° yarns (courtesy of CTMI)

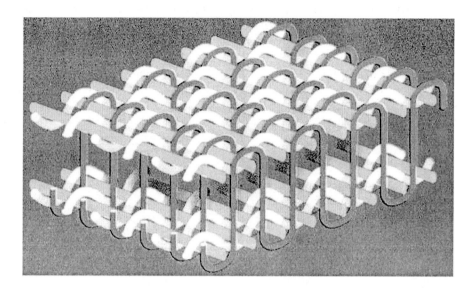

Figure 2.11 Illustration of Distance Fabric material

2.3.1 2D Braiding

The standard 2D braiding technique is illustrated in Figure 2.12, which demonstrates how the counter-rotation of two sets of yarn carriers around a circular frame forms the braided fabric. This movement of the yarn carriers is accomplished through the use of "horn gears" which allow the transfer of the carriers from one gear to the next. The fabric architecture produced by this process is highly interlinked and normally in a flat or tubular form, as shown in Figure 2.13. The style and size of the braided fabric and its production rate are dependent upon a number of variables (Soebroto et al., 1990), amongst which are the number of braiding yarns, their size and the required braid angle. The equations that relate these variables dictate the range of braided fabric that can be produced on any one machine. Generally though, braiding is more suited to the manufacture of narrow width flat or tubular fabric and not as capable as weaving in the production of large volumes of wide fabrics. Typical large braiding machines tend to have 144 yarn carriers, however, larger braiding machines, up to 800 carriers (A&P Technology, 1997), are now coming into commercial operation and this will allow braided fabric to be produced in larger diameters and at a faster throughput.

The braiding process can also be used with mandrels to make quite intricate preform shapes (see Figure 2.14). By suitable design of the mandrel and selection of the braiding parameters, braided fabric can be produced over the top of mandrels that vary in cross-sectional shape or dimension along their length. Attachment points or holes can also be braided into the preform, thus saving extra steps in the component finishing, and improving the mechanical performance of the component by retaining an unbroken fibre reinforcement at the attachment site. Thus, within the limitations of fabric size and production rate, braiding is seen to be a very flexible process in the range of products that are capable of being manufactured. In particular, unlike the standard weaving process, braiding can produce fabric that contains fibres at ±45° (or other angles) as well as 0°, although fibres placed in the 90° direction are not possible with the standard braiding process.

   The primary difficulty with the traditional braiding technique is that it cannot make thick-walled structures unless the mandrel is repeatedly braided over. This can be done but it only produces a multilayer structure without through-thickness reinforcement. To manufacture true three-dimensional braided preforms it was necessary for new braiding techniques to be developed.

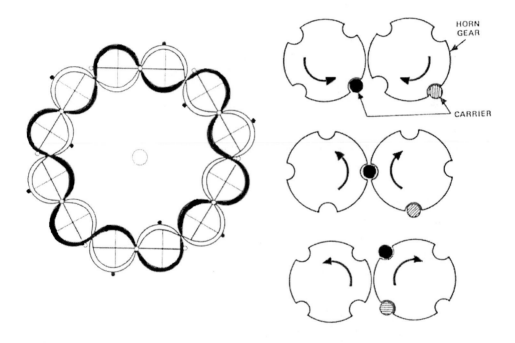

Figure 2.12 Illustration of standard braiding process using horn gears

2.3.2 Four-Step 3D Braiding

The late 1960's saw an interest in the use of three-dimensional braiding to construct carbon/carbon aerospace components and a number of processes were developed to achieve this goal (Ko, 1982; Brown, 1985). One of the first three-dimensional braiding processes (Omniweave) was developed by General Electric (Stover et al., 1971), and further developed and patented by Florentine (1982) under the name of Magnaweave. This process (known as 4-step, or row-and-column) utilises a flat bed containing rows and columns of yarn carriers that form the shape of the required preform (see Figure 2.15). Additional carriers are added to the outside of the array, the precise location and quantity of which depends upon the exact preform shape and structure required. There are four separate sequences of row and column motion, shown in Figure 2.15, which act to interlock the yarns and produce the braided preform. The yarns are mechanically forced into the structure between each step to consolidate the structure in a similar process to the use of a reed in weaving. The motion of the rows, columns and take-up can be altered to obtain preforms with different braid patterns and thus control the mechanical properties of the preform in the three principal directions.

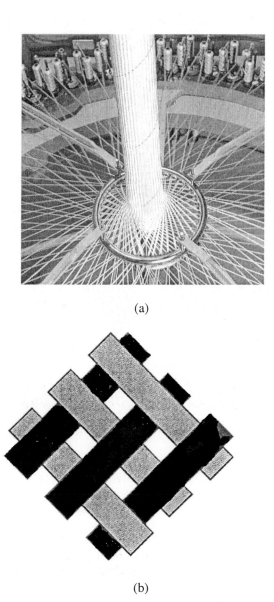

(a)

(b)

Figure 2.13 (a) Production of standard braided tubular fabric, (b) Schematic of typical braid architecture

Braider

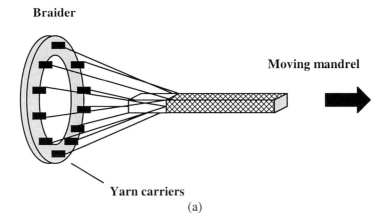

Moving mandrel

Yarn carriers

(a)

(b)

Figure 2.14 (a) Schematic of braiding over a moving mandrel, b) Example of braiding over a T-shaped mandrel (courtesy of the Cooperative Research Centre for Advanced Composite Structures, Ltd)

The process of 4-step braiding can also be accomplished with a cylindrical equipment configuration. An example of this braiding process, called Through-the-Thickness® braiding, was developed at Atlantic Research Corporation (Brown, 1985; 1988). The equipment consists of a number of identical rings situated side by side in an axial arrangement. These rings contain grooves within which the yarn carriers can move from ring to ring in an axial direction. Movement circumferentially is achieved through rotation of the rings, thus accomplishing the 4-step process as shown in Figure 2.16.

This type of cylindrical arrangement has the advantage that it is more efficient with space than the flat-bed arrangement. Both equipment configurations can be easily expanded through the addition of extra rings or flat tiles respectively (Thaxton et al., 1991).

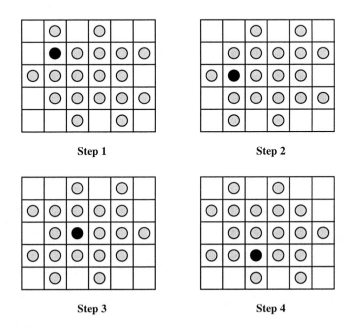

Figure 2.15 Schematic of the 4-Step braiding process

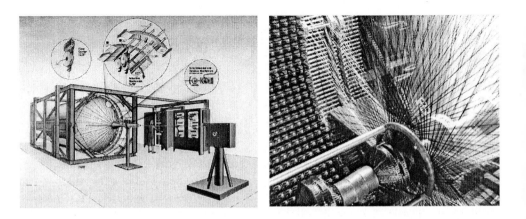

Figure 2.16 Through-the-Thickness® equipment developed at Atlantic Research Corporation (courtesy of Atlantic Research Corporation)

2.3.3 Two-Step 3D Braiding

The second style of flat bed braiding is referred to as 2-step (Popper and McConnell, 1987; Ko et al., 1988; McConnell and Popper, 1988). Unlike the 4-step process, the 2-step includes a large number of yarns fixed in the axial direction and a smaller number of braiding yarns. The arrangement of axial carriers defines the shape of the preform to be braided (see Figure 2.17) and the braiding carriers are distributed around the perimeter of the axial carrier array. The process consists of two steps in which the braiding carriers move completely though the structure between the axial carriers. This relatively simple sequence of motions is capable of forming preforms of essentially any shape, including circular and hollow. The motion also allows the braid to be pulled tight by yarn tension alone and thus the 2-step process does not require mechanical compaction, unlike the 4-step process.

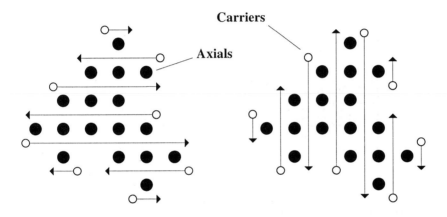

Figure 2.17 Schematic of the 2-Step braiding process

Both the 4-step and the 2-step braiding processes are capable of forming quite intricate shapes as shown schematically in Figure 2.18 (Ko, 1989b) and have been successfully used with a range of fibre materials; glass, carbon, aramid, ceramic and metal. It is possible to braid inserts or holes into the structure that have a greater degree of stability than holes that have been machined. The braid pattern can be varied during operation so that a change in cross-sectional shape is possible, including introducing a taper to the preform. Thick-walled tubular structures can also be made by suitable arrangement of the carriers. Flat preforms can be made from tubular preforms by braiding splits or bifurcations into the preform then cutting and opening it out to the required shape (Brown and Crow, 1992). A bend is also possible as well as a bifurcation, which will allow junctions to be produced and these processes even allow 90° yarns to be laid into the preform during manufacture. Further development of the 2-step and 4-step braiding techniques have concentrated primarily on computer-aided design of the braided preform and improving the process of controlling the transfer of the yarn carrier across the bed (Huey, 1994; Roberts and Douglas, 1995). This includes the use of computer

3D Fibre Reinforced Polymer Composites

controlled horn gears on the flat bed arrangement as shown in Figure 2.19 (Kimbara et al., 1995; Schneider et al., 1998; Laourine et al., 2000).

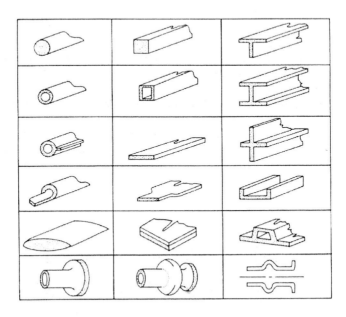

Figure 2.18 Examples of possible 3D braided preforms (Ko, 1989b)

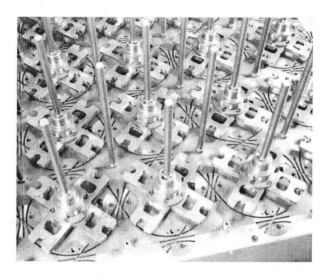

Figure 2.19 Computer controlled horn gears for the transfer of the yarn carrier across a flat bed braider.

2.3.4 Multilayer Interlock Braiding

A different class of three-dimensional braiding does not rely upon the 2-step and 4-step processes previously described, and is considered to be closer to the traditional process of 2D braiding in its operation. This proprietary braiding process, called "multilayer interlock braiding", was developed at Albany International Research Corporation (Brookstein, 1991; Brookstein et al., 1993) and the machinery is analogous to a number of standard circular braiders being joined together to form a cylindrical braiding frame. This frame has a number of parallel braiding tracks around the circumference of the cylinder but the mechanism allows the transfer of yarn carriers between adjacent tracks thus forming a multilayer braided fabric with yarns interlocking adjacent layers (see Figure 2.20). The multilayer interlock braid differs from both the 4-step and 2-step braids in that the interlocking yarns are primarily in the plane of the structure and thus do not significantly reduce the in-plane properties of the preform. The 4-step and 2-step processes produce a greater degree of interlinking as the braiding yarns travel through the thickness of the preform, but therefore contribute less to the in-plane performance of the preform.

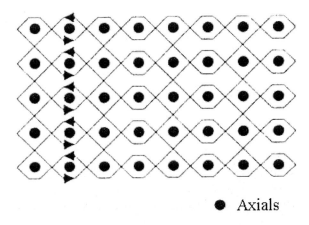

● Axials

Figure 2.20 Schematic of the multilayer interlock braiding process

A disadvantage of the multilayer interlock equipment is that due to the conventional sinusoidal movement of the yarn carriers to form the preform, the equipment is not able to have the density of yarn carriers that is possible with the 2-step and 4-step machines. The consequence of this is that multilayer interlock braiders will be larger than 2-step and 4-step machines for a comparable number of carriers and are considered to be less versatile in the range of preform architectures produced (Kostar and Chou, 1999). However the use of the traditional horn gear mechanisms offers improved braiding speed over the 2-step and 4-step processes.

There are a number of disadvantages with all the 3D braiding processes described here (Kostar and Chou, 1999). Firstly, compared to other textile processes, braiding can only make preforms of small scale relative to the size of the machinery. Also, the length of preform that can be braided before re-supply of the yarn is necessary is limited by the need for the yarn to be on the moving carriers, which ideally must be small and light for rapid braid production. Thus the production of long lengths of the preform can be slow due to the need to re-stock the yarn carriers. One of the greatest current disadvantages however is the fact that the 3D braiding process is still very much at the machinery development stage. Therefore there are limitations to the type of preform that can be made commercially and there are very few companies that have the necessary experience and equipment to manufacture these preforms.

2.4 KNITTING

Knitting may not at first appear to be a manufacturing technique that would be suitable for use in the production of composite components and it is arguably the least used and understood of the four classes of textile processes described here. However, the knitted carbon and glass fabric that can be produced on standard industrial knitting machines has particular properties that potentially make it ideally suited for certain composite components.

2.4.1 Warp and Weft Knitting

Two traditional knitting processes, weft knitting and warp knitting, are available to manufacture preforms for composite structures. Both of these techniques can be performed upon standard, industrial knitting machines with high performance yarns such as glass and carbon. One critical issue that must be considered is that the more advanced knitting machines have electronic control systems close to the knitting region where broken fibres can be generated. The use of carbon yarns with these machines should be avoided as loose carbon fibres can generate electrical shorts. In warp knitting there are multiple yarns being fed into the machine in the direction of fabric production, and each yarn forms a line of knit loops in the fabric direction. For weft knitting there is only a single feed of yarn coming into the machine at 90° to the direction of fabric production and this yarn forms a row of knit loops across the width of the fabric (see Figure 2.21).

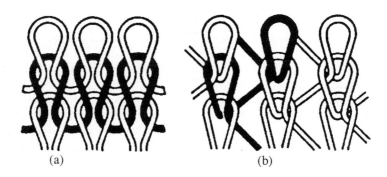

(a)   (b)

Figure 2.21 Illustration of typical a) weft and b) warp knitted fabric architectures

The formation of the knitted fabric is accomplished through a row of closely spaced needles (needle bed) which pull loops of yarn through previously formed knit loops (Figure 2.22). The needle bed can be in a circular or flat configuration and an increase in the number of needle beds available in the machine for knitting increases the potential complexity of the fabric knit architecture. For weft knitted fabrics the motion of the yarn carrier as it travels across the width of the needle bed (or around the circumference for circular machines) draws the yarn into the needles for knitting (Figure 2.23). In much the same way as weaving, warp knitting machines have an individual supply of yarn feeding each knitting needle.

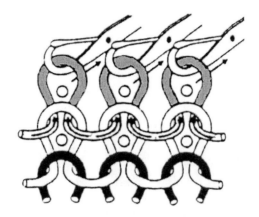

Figure 2.22 Illustration of knitting process

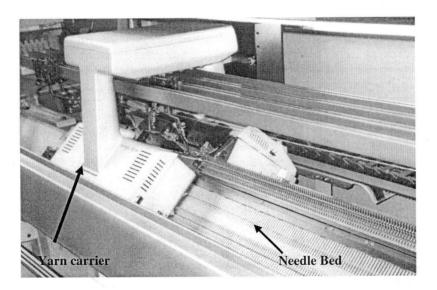

Figure 2.23 Flat bed knitting machine showing the yarn carrier and needle beds

Standard warp and weft knitted fabric are regarded by many as 2D fabric, however, machines with two or more needle beds are capable of producing multilayer fabrics with yarns that traverse between the layers. Figure 2.24 shows a schematic of such a fabric and the range of knit architectures that can be produced with current industrial machines is quite extensive. These flat fabrics can also be formed with variable widths, splits to allow multiple, parallel fabrics to be formed, and holes with sealed edges.

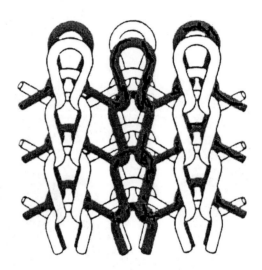

Figure 2.24 Schematic of a multilayer knitted fabric

It is clear from the illustrations of knit architectures that the primary difference between knitted fabric and fabric made by the other textile processes described here is in the high degree of yarn curvature that results from the knitting process. This architecture results in a fabric that will provide less structural strength to a composite (compared to woven and braided fabrics) but is highly conformable and thus ideally suited to manufacture relatively non-structural components of complex shape. This conformability means that layers of knitted fabric can be stretched to cover the complete tool surface without the need to cut and overlap sections. This reduces the amount of material wastage and helps to decrease the costs of manufacturing complex shape components (Bannister and Nicolaidis, 1998). Examples of such components are shown in Figure 2.25.

Changing the knit architecture can vary the properties of knitted fabric itself quite significantly. In this fashion, characteristics such as fabric extensibility, areal weight, thickness, surface texture, etc, can all be controlled quite closely. This allows knitted fabric to be tailor-made to suit the particular component being produced. Both warp and weft knitting also have the ability to produce fabric with relatively straight, oriented sections of the knitting loop (see Figure 2.26) that can be designed to improve the in-plane mechanical performance of the fabric. Warp knitting in particular has been used to produce fabric with additional straight yarns laid into and bound together by the knit structure, but this will be described more fully in a later section.

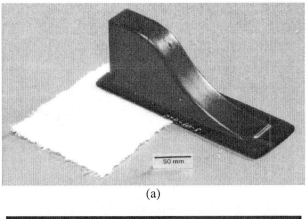

(a)

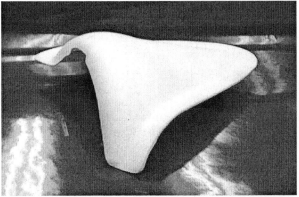

(b)

Figure 2.25 Examples of complex aerospace components manufactured with flat knitted fabric a) Helicopter door track pocket, b) Aircraft push rod fairing (courtesy of the Cooperative Research Centre for Advanced Composite Structures, Ltd)

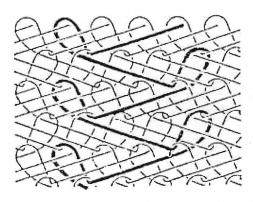

Figure 2.26 Illustration of a warp knitted fabric with oriented sections of yarn.

2.4.2 Three-Dimensional Shaping

As well as producing highly conformable flat fabric, the knitting process can be used to manufacture more complex-shaped items. Since the 1990's significant advances in flat-bed machine technology and design and control software has allowed the development of commercial knitting machines that are capable of forming complex 3D shapes. The leading knitting machine companies, Stoll (Germany) and Shima Seiki (Japan), have lead the research and technical developments in this area and each has commercialised their own machinery capable of producing 3D shapes. The most important developments have been in the use of electronic controls for needle selection and knit loop transfer, and in the sophisticated mechanisms that allow specific areas of the fabric to be held and their movement controlled (Lo, 1999; Editor, 1996; 1997; Reider, 1996; Stoll GmbH, 1999). These developments allow the knit architecture and the way in which the fabric is controlled, to be designed such that as the fabric is manufactured it will form itself into the required three-dimensional preform shape with a minimum of material wastage, examples of which are shown in Figure 2.27. This can be accomplished without fabric overlap or seams and with the fabric properties capable of being designed to be uniform throughout the whole structure. This process is capable of cutting the manufacturing costs for complex-shaped components as the time required to form the component shape would be dramatically reduced when compared to the use of more traditional composite manufacturing techniques (Vuure et al., 1999). In spite of the relative infancy of this area of research a number of net-shaped components have already been demonstrated in high performance yarns including car wheel wells (Vuure et al., 1999), T-pipe junctions, cones, flanged pipes & domes (Epstein and Nurmi, 1991), and jet engine parts (Robinson and Ashton, 1994).

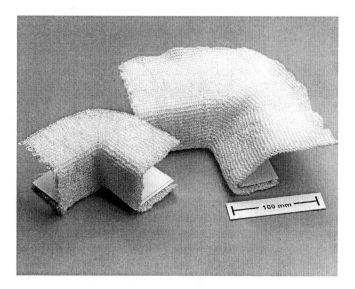

Figure 2.27 Examples of shape knitted corner fabrics designed for composite window frames (courtesy of the Cooperative Research Centre for Advanced Composite Structures, Ltd)

A specialised sub-group of 3D knitted preforms are sandwich fabrics, which were developed by Verpoest et al. in the mid-1990's (Verpoest et al., 1995). They are produced in a similar fashion to 3D woven sandwich fabrics by simultaneously knitting top and bottom skins on a double-bed, warp knitting machine. As the two fabrics are being formed, yarns are swapped between the two faces to create the connecting pile yarns, thus binding the two faces into an integral sandwich fabric. The density of the pile yarns can be varied and their orientation can be aligned vertically or at an angle to the faces in the warp direction. The two needle beds can also be programmed to produce different knit architectures and thus produce face fabrics with different physical characteristics.

As with 3D woven Distance Fabrics, the 3D knitted sandwich fabrics can produce composite sandwich products with high peel and delamination resistance and although their face fabrics will have reduced mechanical performance compared to Distance Fabric faces, their knit architecture allows them to form far more complex shapes than is possible with Distance Fabrics (Verpoest et al., 1995; Mouritz et al., 1999).

2.4.3 Non-Crimp Fabrics

A manufacturing technique that combines aspects of weaving and knitting is known by either of the names; Multi-Axial Warp Knitting or stitch-bonding, but is perhaps most commonly referred to by the style of fabric it produces, Non-Crimp Fabric (NCF). This fabric can be produced with glass, carbon or aramid yarn (or with combinations of these) and is unique in that fabric can contain relatively uncrimped yarns orientated at 0° and at angles that can vary between +20° to -20°. There are a number of generic manufacturing processes which can be employed to produce NCF. The most commonly used process is that developed by the LIBA Machine Company of Germany. A schematic of this process is shown in Figure 2.28 together with an example of the type of fabric that can be produced.

As illustrated in Figure 2.28, yarns are feed from a creel system (1) and are laid onto a long table at the orientations required via placement heads (2), an example of which is shown in Figure 2.29. These placement heads travel across the table and secure the yarns at either side on a chain of needles (3) that travel along the table as the fabric is manufactured. The lay-up of the final fabric is dictated by the control of the placement heads motion. As well as angled fibres, if required, a chopped strand mat can be incorporated into the fabric by the use of a chopper system (4) and further fleeces or mats can be inserted through the use of two roll-carriers (5). The 0° fibres are the last to be placed and can be feed from a beam (6) or a creel system and the multiple layers of the fabric are linked together by a warp knitting machine (7). This machine has specially designed sharp-head needles that are positioned such that the knitting process does not penetrate and damage any yarns but instead forms the knit loop in between the yarns (see Figure 2.30). In current, commercially available fabric the knit thread is normally polyester, but techniques are being developed to manufacture high quality fabric with glass or carbon knitting thread.

The process is flexible in that the variety of lay-ups is dictated only by the number and order of the "stations" (i.e. 90°, 45°, chopped fibre, fleece mats, etc) that are linked together along the length of the production table. However, due to the need to precisely locate the angled yarns on the needle chains and to ensure the knitting needles do not damage the yarns, there are some restrictions on the size of yarns used and the areal weights that can be obtained for each layer of orientated yarn. Also current production machines are only capable of producing fabric with a maximum of 8 layers and the 0° yarns must be placed on an outer layer. However, large widths of fabric can be produced, up to 2.5 m for LIBA machines, and the rate of production is fast, up to 45 linear metres/hour (Kamiya et al., 2000). This makes this production technique highly suited for large volume production.

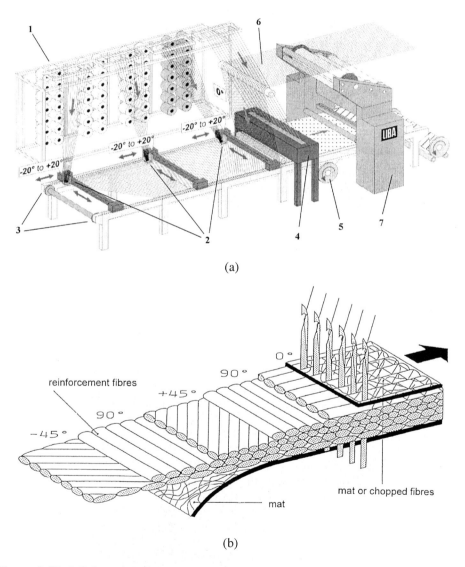

(a)

(b)

Figure 2.28 a) Schematic of the LIBA process for manufacturing noncrimp fabric. 1 = creel system; 2 = placement heads; 3 = needles, 4 = chopper system; 5 = roll carriers; 6 = beam to feed 0° fibres; 7 = warp knitting machine. b) An example of the type of fabric that can be produced with this process (courtesy of LIBA-Maschinenfabrik GmbH)

Figure 2.29 Examples of the fibre placement heads (courtesy of LIBA-Maschinenfabrik GmH)

Figure 2.30 Knit loop formation (courtesy of LIBA-Maschinenfabrik GmbH)

In spite of the restrictions, non-crimp fabric is being used extensively for the manufacture of high performance yachts and in the manufacture of wind turbine blades. Its use is also increasing within the aerospace industry and it is considered to be the prime material candidate for use in future aircraft programs (Hinrichsen, 2000). This fabric has the advantages that fewer numbers of layers need be used to build up the required structure, therefore reducing the cost of labour.   Due to the relatively uncrimped nature of the yarns, laminates produced using NCF have been found to exhibit superior in-plane properties for a given volume fraction of reinforcement than do laminates produced using woven fabric in which the yarns can be more highly crimped (Hogg et al., 1993).   However, unlike the true 3D structures described in earlier sections (weaving, braiding, etc.) the polyester knitting thread does not improve the impact performance of the composite. Non-crimp fabric has also shown a much greater ability to conform to relatively complex shapes without the wrinkling that is normally produced in standard woven fabric. This is due to the ability of the fabric layers to shear a certain amount relative to each other without the knit loops restricting this movement.

2.5 STITCHING

2.5.1 Traditional Stitching

Although the use of stitching in the production of composite components has only been reported since the 1980's, it is arguably the simplest of the four main textile manufacturing techniques that have been described here and one that can be performed with the smallest investment in specialised machinery. Basically the stitching process consists of inserting a needle, carrying the stitch thread, through a stack of fabric layers to form a 3D structure (see Figure 2.31). Standard textile industry stitching equipment is capable of stitching preforms of glass and carbon fabrics and there are many high performance yarns that can be used as stitching threads. Aramid yarns have been the most commonly used for stitched composites as they are relatively easy to use in stitching machines and are more resistant to rough handling than glass and carbon. However the use of aramid stitching threads can cause difficulties in the final composite component due to their propensity to absorb moisture and the difficulty in bonding the aramid yarn to many standard polymer resins. The manufacturer must therefore be aware that these problems may lead to a reduction in the mechanical performance of the component in certain situations. Glass and carbon yarn do not have the problems of moisture absorption and weak interfaces that aramid yarn does, but they are significantly more difficult to use in stitching machines. This is due to their inherent brittleness, which can lead to yarn breakage when stitch knots are being formed and fraying of the yarn in its passage through the stitching machine. Apart from trying to minimise the potential fraying on the stitch thread the main requirement for a suitable stitching machine is that the needle be capable of penetrating through the number of fabric layers to be stitched together in a precise and controlled manner.

Although common, industrial stitching equipment can be used, there has been some development of more complex machines specifically designed for the production of stitched composite components. To date the most ambitious program has been that undertaken by NASA in association with Boeing (Beckworth and Hyland, 1998). This project has developed a 28 metre long stitching machine with the aim to manufacture impact-tolerant composite aircraft wing components that are 25% lighter and 20% cheaper than equivalent aluminium parts. Parts have already been manufactured with this equipment and tested successfully (Phillips, 2000), however the capital costs involved in a stitching machine with these capabilities would be beyond the scope of most composite manufacturers. More recently, machinery advancement has concentrated upon the development of computer-controlled robotic stitching heads that are capable of stitching across a complex, curved surface (Wittig, 2000; Klopp et al., 2000). This equipment is also capable of stitching from one side only (see Figure 2.32), which allows (if required) the stitching step to be done directly on the preform as it sits on the tool surface, an advantage over more common machines which need access to both sides of the preform during the stitching process.

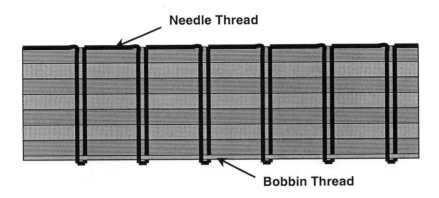

Figure 2.31 Illustration of a stitch pattern through a composite laminate

Stitching has a number of advantages over other textile processes. Firstly, it is possible to stitch both dry and prepreg fabric, although the tackiness of the prepreg makes the process difficult and generally creates more damage within the prepreg material than in the dry fabric. Stitching also utilises the standard two-dimensional fabrics that are commonly in use within the composite industry therefore there is a sense of familiarity concerning the material systems. The use of standard fabric also allows a greater degree of flexibility in the fabric lay-up of the component than is possible with the other textile processes, which have restrictions on the fibre orientations that can be produced. Through the use of robotic mechanisms, it is also possible to automate the stitching of the fabric and thus create a highly automated and economical production process (Bauer, 2000).

Stitching is not restricted to a "global" stitching of the complete component. If required, stitches can be placed only in areas which would benefit from through-thickness reinforcement, such as along the edge of the component or around holes. The density, stitch pattern and thread material can also be varied as required across the component therefore this technique has a great deal of flexibility in the arrangement of the through-thickness reinforcement. Stitching can also be used to construct complex three-dimensional shapes by stitching a number of separate components together (see

Figure 2.33). This not only increases the through-thickness strength of the final component but also produces a net-shape preform that can be handled without fear of fabric distortion.

a)

b)

Figure 2.32 a) Illustration of one-sided stitching technique, b) Example of commercially available robotic, one-sided stitching machine (courtesy of Altin Nähtechnik GmbH)

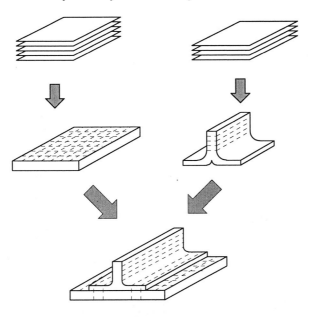

Figure 2.33 Illustration of complex preform manufacture via stitching

There are disadvantages with the stitching process, the main one of which is a reduction of the in-plane properties of the resultant composite component (i.e. tension, compression, shear, etc.). As the needle penetrates the fabric it can cause localised in-plane fibre damage and fabric distortion which has been found to reduce the mechanical performance of the composite (Mouritz et al., 1997; Mouritz and Cox, 2000). This reduction in performance can be aggravated by the surface loop of the stitch, which can also crimp the fabric in the thickness direction if the tension in the stitch thread is high. The presence of the stitch thread and the distortion in the fabric that it creates also causes a resin-rich pocket to be formed within the composite. This pocket can act as a potential crack initiator, which can possibly affect the long-term environmental behaviour of the material. More detail on the damage caused during stitching and the mechanical performance of stitched composites can be found in Chapter 8.

2.5.2 Technical Embroidery

A version of stitching which can be used to provide localised in-plane reinforcement together with through-thickness reinforcement is technical embroidery. In this process a reinforcement yarn is fed into the path of the stitching head and is stitched onto the surface of the preform (see Figure 2.34). With current computer controlled embroidery heads it is possible to accurately place this in-plane yarn in quite complex paths, which allows high stress regions of a component to be reinforced by fibres laid in the maximum stress direction.

Although this technology appears best suited for the placement of localised reinforcement, the technical embroidery technique can also be used to construct complete preforms containing an optimised fibre pattern. Current machinery would tend to limit the size of preforms made in this fashion but the process has the advantage that it is capable of high levels of automation (see Figure 2.35). This manufacturing technique could also be considered a version of the fibre placement technology.

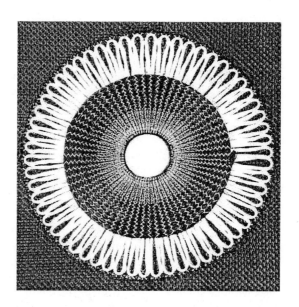

Figure 2.34 Example of a fibre oriented reinforcement manufactured via technical embroidery (courtesy of Hightex GmbH)

Figure 2.35 Multiple preform manufacture with automated technical embroidery equipment (courtesy of Hightex GmbH)

2.5.3 Z-Pinning

An alternate method to the standard stitching process was first described in late 1980's (Evans and Boyce, 1989; Boyce et al., 1989) and subsequently has been commercially developed by the company Aztex (a subsidiary of Foster-Miller) as Z-Fiber™ technology (Freitas et al., 1996). The technology consists of embedding previously cured reinforcement fibres into a thermoplastic foam that is then placed on top of a prepreg, or dry fabric, lay-up and vacuum bagged. Through judicious choice of the material, the foam will collapse as the temperature and pressure are increased, allowing the fibres to be slowly pushed into the lay-up (see Figure 2.36). This method can be used during the normal autoclave cure of prepreg and for both prepreg and dry fabric can be performed whilst the lay-up is on the tool surface itself, thus saving extra steps in the manufacturing process. A version of this technology can be used at room temperatures as it utilises an ultrasonic horn that heats up a local area of the z-pin foam and preform, thus allowing a plunger to push the pre-cured reinforcement yarn into the lay-up. Both methods have been successfully applied to carbon/epoxy composites with silicon carbide, boron and carbon reinforcement yarns. Chapter 9 contains further details on this technology and the mechanical performance of z-pinned composites.

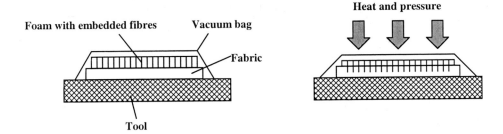

Figure 2.36 Illustration of z-pinning process

2.6 SUMMARY

The four textile processes of stitching, weaving, braiding and knitting, have the potential to significantly reduce the cost of manufacturing many composite components and produce structures that have improved mechanical performance in critical design cases such as impact. Each of these processes has been briefly described here and their advantages and limitations noted. The main aspects of these manufacturing techniques have been summarised in Table 2.1 and reviews of these textile processes can be found in the published literature (Ko, 1989b; Mouritz et al., 1999; Kamiya et al., 2000). However, one manufacturing issue that has only been only briefly mentioned here is the potential of each manufacturing process to cause significant damage to the reinforcement yarns and thus degrade the performance of the final composite. Although this issue as been partly explored for the stitching process (Mouritz et al., 1997; Mouritz and Cox, 2000) very little investigation has been done on the other techniques mentioned here, although recent work has shown that the effects of processing damage can be significant for 3D weaving (Lee et al., in press).

Table 2.1 Description of advanced textile manufacturing techniques

Textile process	Preform style	Fibre orientation	Productivity/setup
Stitching (general)	Complex preforms possible by combining several structures	Dependant upon basic fabric being stitched	High productivity. Short setup time
Stitching (embroidery)	Additional fibres incorporated onto basic fabric	Complex fibre orientations possible, e.g. maximum stress direction	Moderate productivity Short setup time
3D Weaving	Flat fabrics, simple profiles, integral stiffened structures & integral sandwich structures	Wide range of through-thickness architectures possible but in-plane fibres generally limited to 0/90 directions (except with advanced looms)	High productivity Long setup time
3D Braiding	Open & closed profiles (I, C, L, Z, O, T,...) & flat fabrics	0 degree fibres. Braiding fibres between 0-80 degrees. 90 degree fibres possible	Medium productivity Long setup time
Knitting (weft and warp)	Flat fabrics, integral sandwich structures & very complex preforms	Highly looped fibres in mesh-like structure	Medium productivity Short setup time
Knitting (non-crimp)	Flat fabrics	Multi-axial in-plane orientation 0/90/+45/-45. Up to 8 layers possible	High productivity Long setup time

It should be stated that these textile manufacturing techniques will not be applicable for all composite components. Design or manufacturing criteria that favour the use of a particular textile process for one application may not necessarily be relevant for another. It is also possible that for some structures it may be necessary to combine a number of the textile processes in order to obtain a product that satisfies the many, and often conflicting, requirements of cost, performance, production rate, manufacturing risk, etc (Broslus and Clarke, 1991). This intimate connection between the textile manufacturing process, the required preform design, the cost and the performance of the resultant component is of particular importance. It has been mentioned in the descriptions of the various textile processes that there is a very large range of possible preform architectures that can be produced, each with its own mechanical performance and associated cost. It is therefore critical that in the design of any component early consideration is given to the method of manufacture as only slight, relatively unimportant changes to component shape or required performance may result in significant changes to the manufacturing process utilised and the cost of final preform.

In spite of the relative youth of these manufacturing techniques, advanced textile preforms are beginning to be used in the manufacture of composite components (Hranac, 2001). The potential savings in cost and improvements in performance that can be realised through the use of these processes are sufficiently attractive that extensive efforts are being put into further developing these processes. It is not yet clear how far these developments will go, but as designers and manufacturers become more familiar with the advanced textile techniques on offer, the use of these techniques will become more commonplace.

Chapter 3

Preform Consolidation

3.1 INTRODUCTION

The 3D textile preform production techniques outlined in the previous chapter are only the first stage in the production of a 3D fibre reinforced composite material. The use of sophisticated equipment and the intelligent design of the preform will all be to no avail if there is no adequate technology for consolidating the preform with polymer resin.

Some traditional methods of composite material production are simply not suited for use with 3D textile preforms. Hand impregnation involves the use of brushes and rollers to physically work the resin into the fibre preform, which can cause distortion of the preform architecture. It is also not capable of removing all entrapped air from the consolidated composite due to the process being performed at atmospheric pressure. This would result in a component of low quality that would be unsuitable for the high performance tasks normally associated with 3D textile preforms.

The pultrusion process involves a preform being pulled in a continuous fashion through a resin bath in which it is fully wet out. It then travels into a heated die where the resin is cured rapidly and a fully consolidated product emerges from the die where it is cut to the required length. It is theoretically possible to consolidate 3D preforms via the pultrusion process and there would be significant advantages to this as a single source of fabric would be more cost efficient to set up and control compared with the multitude of yarn and 2D fabrics sources that are currently used. However, the current wet out process involves the fabric and yarn having to follow complex paths around guide bars in order to work the resin fully into the fibres. This would severely distort the fibre architecture of a 3D preform thus compromising the mechanical performance of the final composite part.

The use of commingled yarns to produce the preform is another possible consolidation route. These yarns consist of the reinforcement fibres intermingled with fibres of thermoplastic resin or particles of partially cured thermosetting resin. These commingled yarns can then be processed into textile preforms via the techniques outlined in Chapter 2, although for commingled thermoset yarns it is more difficult as the yarns often become less flexible through the commingling process. The application of heat and pressure then causes the resin to melt and wet out the preform. The difficulty arises in that the volume occupied by the resin relative to the total unconsolidated preform volume is low. Therefore to ensure that the available resin completely fills the fibre reinforcement and the volume fraction of reinforcement fibres is structurally significant, the preform must be dramatically reduced in volume during consolidation. This is generally not a problem for two dimensional fibre architectures as the thickness can be reduced without disrupting the architecture however a three-dimensional fibre architecture will be severely distorted through this consolidation thus rendering this manufacturing route unsuitable.

To date the only general manufacturing process that has been used successfully with 3D fibre preforms is Liquid Moulding (also known as Liquid Composite Moulding). There are many different variations of Liquid Moulding (LM) and the main techniques will be reviewed here. However, there are many issues involved in the successful consolidation of 3D fibre preforms and this chapter can only briefly outline these issues. For a more detailed explanation the reader is referred to publications such as Kruckenberg and Paton (1998), Parnas (2000) and Potter (1997).

3.2 LIQUID MOULDING TECHNIQUES

Within the published literature you will find many variations on the theme of liquid moulding, each with it's process distinctions that, in the eyes of it's developers, differentiate their technique from others and thus make it deserving of its own acronym. In reality, there are three primary liquid moulding techniques from which the other processes are derived.

3.2.1 Resin Transfer Moulding

Resin Transfer Moulding (RTM) is the most commonly used of the three main processes, particularly for the production of high performance aerospace components. The main aspect of this moulding technique which differentiates it from the following two processes is the general direction of flow the resin takes as it infiltrates the preform.

The RTM process is characterised by a primarily in-plane flow of the resin through the preform. The resin is driven into the preform by the pressure of a pump. For very thick or complex shaped parts there will be an element of through-thickness resin flow but essentially the movement of the resin is within the plane of the preform. Figure 3.1 illustrates this basic concept of the RTM process. The in-plane resin flow patterns that can occur within the preform are dictated by the design of the resin inlet and outlet gates. The maximum injection length of the resin into the preform is therefore limited by the in-plane preform permeability, the resin viscosity, the differential pressure driving the resin flow and the rate at which the resin is polymerising. These factors can be quite variable amongst the range of RTM products being produced and the resin systems used in their manufacture but, typically, injection lengths can range up to 2 metres (Räckers, 1998). Higher permeability, lower resin viscosity, higher injection pressures and slower resin cure rate will all act to increase the injection length and thus the size of the part that can be produced. Production of a component larger than the maximum injection length can be accomplished through the use of multiple resin inlet and exit ports therefore one of the main issues which can restrict the size of component produced via RTM is the tooling used in the process.

The tooling used for RTM is most often a closed mould system, thus has two main tools that enclose the preform. This can allow excellent surface finishes and close dimensional tolerances to be obtained if high quality (and normally expensive) tooling materials are used. Heating and cooling systems can also be built into the mould tools to minimise delays in obtaining the required tool temperature. The RTM process usually achieves the high fibre volume fraction of 55-60% normally required in high performance components as the use of quality tooling materials and presses allows for the application of large compaction pressures. This need for, often, expensive tooling and presses with sufficient load capacity limits the size of component that can be economically produced via the RTM process. Cheaper tooling can be used but this often restricts the compaction pressure that can be applied and can potentially reduce the surface quality. These and other tooling issues are discussed further in Section 3.6.

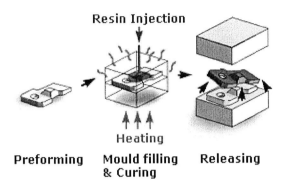

Figure 3.1 Schematic of the RTM process

There are a number of liquid moulding processes related to RTM. Vacuum Assisted RTM (VARTM) is the same as RTM except that vacuum is applied to preform. This aids in consolidation quality through removal of air and speeds up the resin infiltration through an increased pressure differential. Structural Reaction Injection Moulding (SRIM) is similar to RTM and is used primarily in the automotive industry. The main difference is that much higher injection pressures are used to fill the preform quickly as the resin systems are generally fast curing and short cycle times are crucial in automotive production.

3.2.2 Resin Film Infusion

The process of Resin Film Infusion (RFI) is different from the RTM technique in two ways. Firstly, as the name suggests, the resin is initially present within the process as a film rather than a liquid. Secondly, the movement of the resin after heat and pressure is applied and the film melts, is in the thickness direction of the preform not in the plane of the preform as in the RTM process. The essential details of this technique are shown in Figure 3.2. In the RFI process the resin film is placed against the prepared tool surface, covering the necessary part surface area, and the preform is placed on top of the film. A release film, to aid in part removal, and a breather material, to enable the generation of vacuum within the bagged area, is then laid on top of the preform. This lay-up is then bagged in a similar process to prepreg components and can be heated within an oven or autoclave, depending upon the requirement for externally applied pressure. The molten resin is sucked into the fibre preform through capillary effects and the careful placement of vacuum outlets. External pressure can be used to compact the preform to the required fibre volume fraction and also add to the pressure that is forcing the resin to flow.

An advantage of the RFI technique is that, in a similar fashion to prepreg manufacture, only one major tool is needed in the process. For complex parts caul plates and small tools to aid in the compaction of specific areas are often used, however the tooling costs associated with RFI are generally much lower than with RTM.

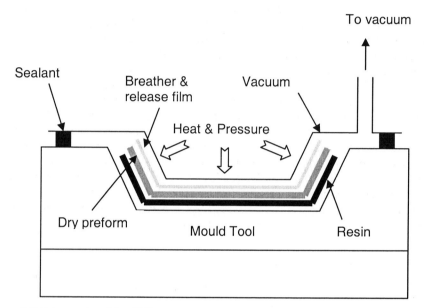

Figure 3.2 Schematic of the RFI process

As the resin flow in RFI is in the thickness direction of the preform, there are not the same part size limitations due to maximum injection length as is the case with RTM. The main criterion in the RFI process is that the resin be capable of flowing through the complete thickness of the preform. This can be a significant issue in the design of the RFI process as many components that are potential candidates tend to be integrally stiffened structures and the height of the stiffener must not be beyond the infusion capability of the resin. The RFI process is therefore more suited for relatively flat, large surface area components whilst the RTM process is used more often for smaller, thicker and more complex parts.

The disadvantages of the RFI process relate to the resin film itself. The manufacture of a resin film suitable for RFI can be quite costly and the price of such a film can be up to twice that of the pure resin (Räckers 1998). A further disadvantage is that the films are quite difficult to handle due to their lack of any supporting carrier material which other film materials have, such as adhesive films. RFI films are also generally of low areal weight so many plies of film must often be stacked together in order to infuse the component. This will increase the labour costs associated in its production.

There do not appear to be any other processing techniques, related to RFI, that are known under different names. The main variation within the RFI process appears to be whether the infusion is conducted in an oven under vacuum pressure or in an autoclave with an additional autoclave-generated pressure.

3.2.3 SCRIMP-based Techniques

The Seemann Composite Resin Infusion Process (SCRIMP) and similar techniques are essentially a mixture of the RTM and RFI processes. Like the RTM process, SCRIMP introduces liquid resin from an external source into the part via a resin inlet port. However, in a similar fashion to RFI, the primary resin flow direction is through the thickness of the preform. This style of resin flow is accomplished through the use of a resin distribution medium which allows the resin to flow quickly over the surface area of the part as it is also infusing through the preform thickness.

Figure 3.3 illustrates the typical set-up of a SCRIMP process. In a similar fashion to RFI, the fibre preform (and any core materials and inserts that may be required) are placed onto a tool together with a resin distribution medium and sealed with a vacuum bag in the conventional way. The part is then placed under vacuum and the resin introduced into the preform through a resin inlet port. The resin is distributed throughout the part via the flow medium and, if required, a series of channels. These channels can be piping on top of the distribution medium or can be channels cut into any core material present. The pressure differential provides the driving force for infusing the resin into the preform, in effect sucking the resin into the preform from the resin container, therefore injection equipment is not required for this process.

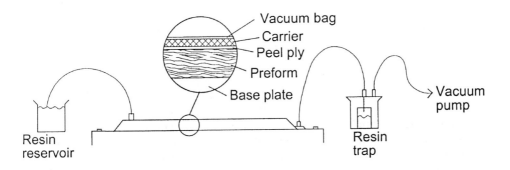

Figure 3.3 Schematic of the SCRIMP process

Like the RFI and conventional prepreg techniques only one tool is needed for the SCRIMP process and thus the tooling costs are significantly less than RTM. However, it also has an advantage over RFI in that raw material costs are reduced due to the use of the cheaper liquid resin rather than the more expensive resin film.

The SCRIMP process has generally become associated with the production of non-aerospace components such as yacht hulls, bus body-shells, refrigerated rail cars, wind turbine blades, etc, as the use of only vacuum pressure to consolidate the preform generally produces components with lower fibre volume fractions than RTM or RFI. Through the careful selection of the resin systems, cure times can be lengthened to the point that very large structures can be economically produced via the SCRIMP technique and yacht hulls of up to 37.5 metres (123 ft) have been manufactured (Stewart 2001).

The SCRIMP process has also been described under a number of other acronyms, VIP (Vacuum Infusion Process) and VBRI (Vacuum Bag Resin Infusion) to name just two. The only apparent differences between all the SCRIMP-based processes appear to relate to the techniques or materials used to distribute the resin rapidly across the surface area of the preform.

3.3 INJECTION EQUIPMENT

Out of the three primary techniques of liquid moulding, two of the processes (RFI and SCRIMP) do not require specialised injection equipment to introduce the resin into the preform. The selection and use of resin injection equipment, as described in this section, is therefore related specifically to the RTM process.

All injection equipment consists of three basic components: the resin storage area, the resin feed apparatus and the delivery hose (an example of an RTM injection machine is shown in Figure 3.4). There are many variations in style and operation of these components that are available through the numerous manufacturers of injection equipment, however one of the first equipment choices that has to be made is influenced by the choice of resin and its handling. Essentially, resins can be handled as either one-part, pre-mixed resin/hardener systems that are injected into the mould via a single valve, or with the resin and hardener kept separate in individual reservoirs and mixed during the injection process in a multi-valve machine.

Both options have their advantages and disadvantages. In the one-part, single valve process, uneven mixing can be eliminated as all the resin components are pre-mixed prior to use. The cure process can also be easier to control as all the resin components have been mixed together at the same time. There are generally less moving parts on single valve machines therefore maintenance can be reduced and the system heating is simplified as only one reservoir is used. Cleaning of the system is generally simpler than multi-valve machines therefore the use of single valve machines is more suited to low production volumes or when a variety of different resin systems are to be used. This is generally seen in the aerospace industry or for research and development. The main disadvantage is that as the resin is pre-mixed it can be curing within the reservoir. Therefore, if too much resin is mixed or delays occur in production, there is a risk that the usable life of the resin will be exceeded and the excess will be wasted.

The main advantage of multi-valve machines is due to the fact that the resin components are kept separate and thus unmixed. This means that the usable life of the resin system is extended and therefore larger volumes of materials can be stored in the reservoirs. As mixing and injection of only the required amount of resin is accomplished, waste is generally reduced. This equipment is most often used in a production-line format where a limited number of resin types are used and the throughput of resin is large, for example, the automotive or sporting goods industries. The disadvantages of the multi-valve machines are the greater degree of maintenance required of the more complex equipment and the possibility of uneven catalysation if incorrect mixing occurs during impregnation.

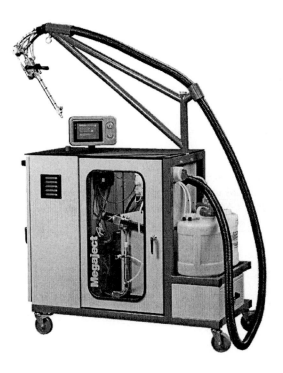

Figure 3.4 Megaject RTM-Pro injection machine (photo courtesy of Plastech T.T. Ltd, UK)

The other main decision affecting the choice of equipment is related to the mechanism used to inject the resin into the mould and here the choice is between equipment designed to produce a constant injection pressure or a constant flow rate.

When constant pressure is being used then the flow rate will vary during impregnation and will decrease with increasing distance from the injection port. The main advantage of this system is the complete control over pressure during injection. This system has the obvious disadvantage of limited control over the resin flow rate, but also the added disadvantages of the pressurisation being limited by available equipment, for example shop air pressure, and the fact that the resin must be held in a pressurisable container, which can limit the volume of resin that can be handled. A heated pressure pot is an example of a constant pressure system.

Constant flow rate systems are usually driven by reciprocating piston pumps and enable repeatable injection times to be maintained, which is important for production-line manufacturing. With these pumps flow is actually semi-constant as flow of a set volume of resin occurs during the piston downstroke and it is stopped during the piston upstroke when resin is drawn in. The advantages of this system include flow rate control, larger resin reservoirs and the ability to have higher injection pressures. The main disadvantage comes from the increasing backpressure generated as the resin flows through the preform. If this pressure becomes too high then the preform can be displaced within the mould and can even cause mould deflection and damage to the equipment. If the flow front is also moving too rapidly in the preform then void entrapment can result, leading to a poor quality part.

Fortunately, many suppliers of injection equipment can now supply systems capable of control over both the flow rate and pressure and contact details for some of these companies are given in Table 3.1.

Table 3.1 Examples of RTM injection equipment suppliers

Aplicator System AB	Metallvä 3, S-435 33, Mölnlycke, Sweden, Tel: +46-31-750-30-00, Fax: +46-31-750-30-01, www.aplicator.se
Ashby Cross Company Inc.	28 Parker Street, Newburyport, MA, 01950, USA, Tel: +1-978-463-0202, Fax: +1-978-463-0505, www.ashbycross.thomasregister.com
Liquid Control Corp.	8400 Port Jackson Ave N.W., N. Canton, Ohio, 44720, USA, Tel: +1-330-494-1313, Fax: +1-330-494-5383, www.liquidcontrol.com
Magnum Venus	1862 Ives Ave, Kent, WA, 98032, USA, Tel: +1-253-854-2660, Fax: +1-253-852-0294, www.venusmagnum.com
Plastech T.T. Ltd.	Unit 1 Delaware Road, Gunnislake, Cornwall, P218 9AR, UK, Tel: +44-1822-832-621, Fax: +44-1822-833-999, www.plastech.co.uk
Radius Engineering Inc.	3474 South 2300 East, Salt Lake City, Utah, 84109, USA, Tel: +1-801-277-2624, Fax: +1-801-277-7232, www.radiuseng.com
Wolfangel GmbH	Roenstgenstr 31, D-71254, Ditzingen, Germany, Tel: +49-07152-51071, Fax: +49-07152-58195, www.wolfangel.com

3.4 RESIN SELECTION

The selection of a resin system for the liquid moulding of 3D fibre preforms is influenced both by the requirements dictated by the use of the composite component and the requirements driven by the manufacturing process. In the first case, the intended application of the composite component will influence the selection of the resin system based upon factors such as mechanical properties, environmental resistance, cost, etc. Although these are important criteria for any resin, they do not directly effect the ability of the resin to be processed under liquid moulding conditions. There are essentially two processing factors that are critical to know in selecting a resin system for successful liquid moulding and these are the resin viscosity and pot life.

The viscosity of the resin must remain low enough during the entire moulding process in order to enable the resin to successfully infuse the complete volume of preform without the need for excessive driving pressures to be used. Within the three types of liquid moulding processes described here, the driving pressure can range from less than 100kPa up to approximately 700kPa which is commonly used in rapid injection processes within the automotive industry. The preform volume fraction and its size also plays a part in determining the necessary resin viscosity, with low fibre volume fractions having a greater permeability to the resin than high volume fraction preforms. However, within the range of injection pressures, preform volume fractions and component sizes, the general rule-of-thumb used is that resins suitable for liquid moulding should have viscosities no higher than 500 cps (centipoise) during moulding. This is particularly true for the high volume fraction preforms used in the aerospace industry as the use of resin systems with viscosities higher than this tends to lead to mould pressures that are difficult to handle and often produces composites with poor fibre impregnation.

Given the critical influence of resin viscosity to the liquid moulding process, the practical definition of resin pot life within liquid moulding is normally defined as the time it takes for the resin system's viscosity to reach a level which prevents further liquid moulding from occurring (generally 500 cps). Depending upon the size and complexity of the part, resin pot lives may be required to run from minutes, for the rapid production of automotive parts, to hours for large marine structures. The time required to fill a preform can be determined from Darcy's Law which relates the flow rate of a resin to parameters such as its viscosity and the preform permeability.

$$\text{Flow rate} = \frac{\text{permeability} \times \text{cross} - \text{sectional area}}{\text{resin viscosity}} \times \frac{\text{pressure drop}}{\text{unit length}}$$

As a significant proportion of liquid moulding processes occur with thermosetting resins, the operator must be aware that the resin will generally be curing throughout the process and thus its viscosity will be increasing with time. The temperature of the resin during liquid moulding will also affect the resin viscosity. The initial viscosity will decrease with increased temperature but the rate of cure will increase, therefore the operator needs to obtain a balance between moulding temperature and pot life in order to ensure that the preform is successfully consolidated. An illustration of how temperature and time affects the resin viscosity for epoxy systems is shown in Figure 3.5

There are a wide variety of resin systems that can be used for liquid moulding and more detailed information can be found in the references Kruckenberg and Paton

(1998), Parnas (2000) and Potter (1997) or directly from resin suppliers such as Hexcel, 3M, Dow Chemical, Bayer, Shell, etc.

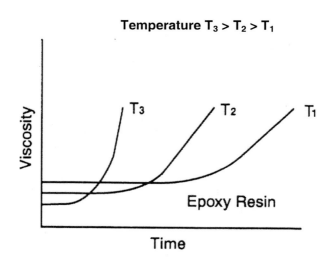

Figure 3.5 Illustration of resin viscosity versus time

3.5 PREFORM CONSIDERATIONS

When liquid moulding is used to consolidate preforms constructed from 2D fabric, one of the most important considerations is the need to keep the preform stable through a means of binding the fabric layers together. Normally this is accomplished by the use of a relatively small amount of binder resin which will be compatible with the matrix resin.

The use of 3D fibre preforms negates the need to use a binder resin as the 3D nature of the fibre architecture creates an inherently stable preform. This is a major advantage of these preforms over those produced from 2D fabric and can lead to significant cost advantages when liquid moulding complex structures (Broslus 1991).

There are however some issues related to the liquid moulding of 3D fibre preforms. Generally the preforms are not produced at the final fibre volume fractions required in the composite structure thus pressure is often used to compact the preform to the required fibre volume fraction. In 2D fabric preforms this is generally not a concern as the pressure is usually applied normal to the fabric layers and thus does not affect the fibre directions. However, with 3D fibre preforms not all the reinforcement will be perpendicular to the pressure therefore the use of compaction pressure can lead to a distortion of the 3D fibre architecture and thus a potential degradation of the composite properties. An allowance for this possible distortion must therefore be made when designing the preform architecture.

A further issue is the potential of having preferential flow directions within the preform that can prevent correct filling with the liquid resin. Many 3D preforms, particularly non-crimp fabrics and those produced by weaving, can have planes of very straight reinforcement in specific directions. This directionality can result in significant differences in preform permeability with orientation that could lead to the resin flowing more swiftly in certain directions ("racetracking") and trapping off unfilled sections of the preform. Accurate knowledge of the preform permeability with orientation and correct design of the liquid moulding process will allow this issue to be overcome.

3.6 TOOLING

The proper design and manufacture of tooling for liquid moulding is a critical part of successfully consolidating a 3D fibre preform. Of the three liquid moulding processes described in this chapter (RTM, RFI and SCRIMP), both RFI and SCRIMP utilise single-sided tools whilst the RTM process requires a closed mould system. Although this difference does allow a greater ability for the RTM process to incorporate integral heating and cooling systems into the tooling, many of the liquid moulding tooling issues are common to all three process styles.

3.6.1 Tool Materials

Generally the first decision that is made in the tool design process is to choose the material from which to manufacture the liquid moulding tool. There are many materials which can be used, ranging from metal (steel, Al, etc) to cast resin, wood or plaster. The choice of material will be influenced by many factors and detailed discussion of these can be found in references such as Potter (1997) and Wadsworth (1998). Some of the primary factors will be briefly discussed here.

The production rate is often one of the most important factors in the selection of tool material. For low volume (100's of parts) or prototype production, composite, cast resin, wood or plaster tools are often used and have the advantage that they are significantly cheaper than metal tools and thus are more suited to low production volumes. For higher production volumes (1,000 – 10,000 + parts), metal tools (steel, aluminium, electroformed nickel, etc) are the only possible choice due to their durability. Although metal tools are more costly on a direct comparison with non-metal, the higher initial tooling costs are generally outweighed by the reduced need to repair or replace them, which is an important consideration in high volume and production rate environments.

The processing conditions and required surface finish also affect the material choice. Metal tools are capable of withstanding far higher service temperatures than non-metal tools and are thus more suited for processes using resins with high cure temperatures. Properly maintained metal tools also produce better surface finishes than non-metal, which is particularly important in industries such as the automotive. Other issues such as the heat transfer requirements and the need for dimensional control can also influence the choice of tool material but generally these are secondary to the issues mentioned above.

3.6.2 Heating and Cooling

The SCRIMP and RFI processes both operate with single-sided tooling therefore heating is generally conducted via an external source such as an autoclave, air convection oven or radiant heaters, or even through the use of electric heating blankets. The selection of a heating system will be dependant upon the size of the part being produced and the processing conditions (heating rates, cure temperature, etc). Generally though the tools are not integrally heated as it is a less efficient, and often more costly, way of applying thermal energy to the preform and resin with a single-sided tool. Cooling for these processes would generally occur via natural cooling in the air.

As the RTM process uses double-sided tooling, integral heating becomes a more likely candidate as a means to apply thermal energy. Normally the mould is heated and cooled using temperature controlled water or oil, although electrical elements can also be used for heating. The mould is constructed with interior channels through which the heating/cooling fluid flows and this normally results in a very efficient, controlled process for heating and cooling the mould. The selection of fluid temperatures will depend upon the required heating rates and cure temperatures but also upon the size of the mould and the thermal properties of the mould material itself. Alternate heating techniques for the RTM process include heated platens in a press, which also has the advantage of providing the mould clamping pressure, and external sources such as ovens. These techniques are normally not as efficient as the integral heating process.

3.6.3 Resin Injection and Venting

This part of the mould design is one of the most critical and, although the exact details of resin flow are different between RTM, RFI and SCRIMP, this issue is relevant to all three of the liquid moulding techniques.

The injection ports (resin inlets) and vents (resin outlets) must be correctly positioned so that the resin will contact all of the preform during its flow. Bypass of any part of the preform will result in dry patches, one of the types of defects that will be discussed in a later section. The factor common to many successful inlet/outlet designs is that the flow path should be arranged such that the resin is flowing into a configuration with decreasing volume. Thus the volume of air left in the preform will be decreasing and this reduction effect helps sweep the air out of the part. Figure 3.6 illustrates examples of good and bad inlet/outlet designs with regard to this rule-of-thumb. The reverse arrangement can be used but this generally requires a greater understanding of the likely resin flow in order to obtain fully wet-out components. Flow modelling can be a very important process to undertake when designing a mould, particularly when the preform permeability is very anisotropic. There are various commercially available software packages that can be used for this task. The details of modelling the flow of resins in liquid moulding processes is explained in greater detail in Parnas (2000).

Vents should be placed so as to draw the resin through preform sections that are difficult to wet out and this is usually at the extreme end of flow paths or dead ends, where the resin will not flow by itself. Vents must also be capable of being individually sealed after the resin begins to bleed out as this will force the resin to flow into other sections of the preform and, when all are sealed, will allow the final curing process to occur under pressure. This will help reduce the possibility of voids in the finished part.

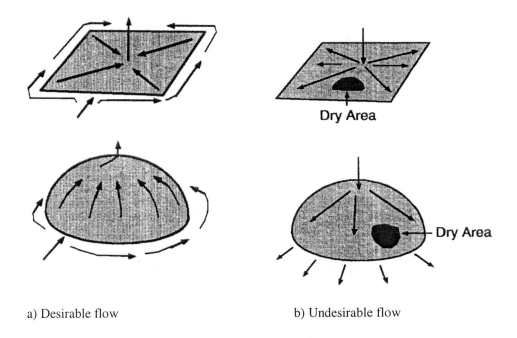

a) Desirable flow   b) Undesirable flow

Figure 3.6 Desirable and undesirable resin flow paths

3.6.4 Sealing

Adequate sealing of the mould is essential if parts of low void content are to be produced. In RFI and SCRIMP this is accomplished with sealant tape in the normal fashion when sealing the preform under a vacuum bag. In RTM the most common way to produce a seal is through the use of elastomeric O-rings (materials such as silicone rubber or Viton®). These O-rings sit in a machined groove within one half of the mould and are compressed when the mould closes. The choice of O-ring material depends upon the required pressure sealing capacity and the maximum temperature it will see during the moulding cycle.

Another method that can be used is the pinch seal technique. Here the preform itself is clamped tightly between the two mould halves to create a region of very high fibre volume fraction. This will increase the resistance to flow of the resin in this area and ideally create an area through which the resin cannot flow over the course of the injection and cure. In reality pinch seals generally allow resin to leak through, which can be a health and safety concern. The final consolidated part will also need more extensive trimming than one produced with an O-ring seal.

3.7 COMPONENT QUALITY

There are a number of factors that can define the quality of the component produced via a liquid moulding process. Factors such as adequate fibre volume fraction, correct fibre orientation, degree of resin cure and interfacial bonding between the fibre and resin are important but are generally controlled through the preform design, resin selection and control over the cure cycle. The primary component quality factor that is a direct result from the process of liquid moulding is the presence of defects such as voids, porosity or dry patches within the component. A dry spot is defined as a region of the preform that has not been wet out by resin, an example of which is shown in Figure 3.7a. Voids (Figure 3.7b) are bubbles of air or other gases, whilst porosity is a collection of voids within a region.

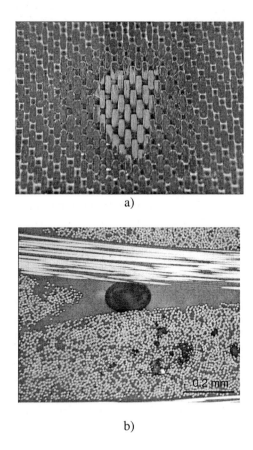

a)

b)

Figure 3.7 a) Typical dry patch b) Typical void between reinforcement tows

Dry spots essentially occur due to the resin not flowing correctly to every part of the preform. This can be due to a poor design of the resin inlet and outlet positions which is often influenced by the complex geometries of the preforms that are generally consolidated by liquid moulding. Dry spots can also be produced by a variation in preform permeability causing unintended resin flow paths. This is particularly apparent at the edges of preforms, where racetracking can often occur, or if the preform contains areas of highly anisotropic permeability that can force the resin flow into unintended directions. It is possible to repair some dry spots after manufacture by a local injection of resin although generally not all of it is eliminated and the area is often weaker. If a dry spot is observed to be forming during moulding a process of repeatedly blocking and opening the outlet whilst keeping the injection pressure on can act to move the trapped air to the outlet due to the varying pressure differentials. This process is known as "burping" (Räckers, Howe & Kruckenberg, 1998).

Voids and porosity can be formed through a number of mechanisms. Air leaks in vacuum assisted liquid moulding can cause large, irregular voids to form and are generally located near the perimeter of the part or near the inlet/outlet positions. They are formed through inadequate sealing. Regular vacuum checks and replacement of seals and fittings can eliminate these defects. Volatiles formed during the resin infusion and cure process can also form gas-filled voids. These are generally observed as small, isolated voids spread evenly through the component. A change of resin type to a non-volatile producing species, or an adjustment of the vacuum pressures or cure temperatures can help eliminate these voids. If this is not possible then an increase in the injection pressure can help minimise the void size. The final void formation mechanism is the mechanical entrapment of air. This is essentially a smaller version of the dry spot formation on the scale of individual tows. The resin contains two paths within the preform through which it can flow, between the tows and within the tows. The driving forces for the two paths are different, capillary action within the tow and the overall hydrodynamic pressure between the tows. If these driving forces are not similar then the flow front can lead in one of the flow paths and lead to the entrapment of air within the other flow path. To overcome this problem vacuum is normally applied in the moulding process therefore any trapped voids will have an internal pressure close to vacuum. As the hydrostatic pressure increases the voids must shrink and may completely collapse.

3.8 SUMMARY

Liquid moulding processes are currently the only viable techniques that can be used to successfully consolidate 3D fibre preforms. There are many different styles of liquid moulding but they all essentially fall into the main groupings of Resin Transfer Moulding (RTM), Resin Film Infusion (RFI) and the Seemann Composite Resin Infusion Process (SCRIMP). A basic description of the various techniques and issues related to the choice of equipment, resin selection, tool design and part quality have been given in this Chapter.

Liquid moulding is in common use within a wide range of industries and is a well established manufacturing process, but is primarily used with traditional 2D fibre reinforcements. The consolidation of 3D preforms via liquid moulding does not appear to hold significant challenges and examples of the use of liquid moulding 3D reinforced composites for commercial and prototype use have been given in Chapter 1. The main deterrent to the increased use of liquid moulded 3D composites appears to be more related to the current ability to manufacture the desired preforms and uncertainty over their mechanical performance.

Chapter 4

Micromechanics Models for Mechanical Properties

4.1 INTRODUCTION

Composite materials are composed of at least two constituent phases, such as fibres and matrix, and their overall behaviour is dependent on the mechanical properties of the constituent phases and the detailed forms in which they interact. Composite materials are intrinsically heterogeneous at the micro-scale. However, the heterogeneous structures of composite materials may be idealised as a homogeneous medium with effective anisotropic properties at the macro-scale, which have been widely and successfully used in practical structural engineering. The procedure to determine the effective properties of a representative volume of composite materials from the known properties of the individual constituents and their detailed interaction is referred to as micromechanics analysis or characterisation of composite materials in a more general sense.

The effective overall behaviours of a composite material are dependant on the mechanical properties of the individual constituents, and their detailed interaction, such as relative volumetric ratios and micro-structural distribution of the individual constituents. While it may be relatively easy to determine the mechanical properties of the individual constituents, comprehensive understanding and accurate consideration of the interaction between the individual constituents at the micro-scale is of vital importance and represents a great challenge in micromechanics analysis. Over the past five decades, many researchers have devoted their efforts to the development of micromechanics analysis techniques to predict mechanical properties of composite materials. Treatment of microstructures and their influence in a composite material is one of the most important efforts. Micromechanics models have been developed to evaluate some effective linear properties of certain composite materials by completely ignoring the influence of microstructures of all constituents. For composite materials with their microstructures having stochastic and probabilistic features, uncertainties of some effective properties can be estimated by determining their corresponding upper and lower bounds. Recently, with the advance of computing and measurement technologies, more accurate evaluation of effective properties for a composite material can be achieved with the aid of more available information on microstructures of all constituents.

This chapter will focus on micromechanics analysis of fibre reinforced composite materials, particularly those materials reinforced long fibres. Typical examples include unidirectional fibre reinforced composites, 2D woven composites as well as 3D fibre reinforced composites. Earlier researches were conducted based on a large number of assumptions for simplifying the analysis procedures. The relevant approaches include rules of mixture approximations, composite cylinder models and the variation boundary method. These methods provide approximate estimation of mechanical properties, but cannot be used to analyse variations of mechanical properties with some important architecture parameters due to the introduction of oversimplified assumptions. On the other hand, it is almost impractical to investigate experimentally the mechanical properties of textile composites and their dependence on the major architecture parameters because of their complexity in geometry and spatial organisation. Hence, it is desirable to develop an analytical approach which is capable of modelling textile composites at a micro geometry level, and predicting effectively the mechanical properties and their dependence on major architecture parameters.

Finite element analysis (FEA) is a useful and versatile approach used by many researchers to predict mechanical properties of composite materials. A number of FEA models have been developed to evaluate the effects of various fibre architecture parameters on the mechanical properties of textile composites.

It is not intended to present or even review in this book all micromechanics methods that have been used or may be potentially useful for characterising 3D fibre reinforced composite materials. Instead, this book aims to provide a brief description of selected micromechanics modelling methods that have been proved to be useful for predicting the in-plane mechanical properties of 3D composites.

4.2 FUNDAMENTALS IN MICROMECHANICS

4.2.1 Generalized Hooke's Law

For an elastic anisotropic material, the generalized Hooke's law is the linear stress-strain relation as given by:

$$\{\sigma\} = [C]\{\varepsilon\} \tag{4.1}$$

where

$$\{\sigma\} = \{\sigma_{11} \quad \sigma_{22} \quad \sigma_{33} \quad \sigma_{23} \quad \sigma_{31} \quad \sigma_{12}\}^T$$

$$\{\varepsilon\} = \{\varepsilon_{11} \quad \varepsilon_{22} \quad \varepsilon_{33} \quad \varepsilon_{23} \quad \varepsilon_{31} \quad \varepsilon_{12}\}^T$$

$$[C] = \begin{bmatrix} C_{11} & C_{12} & C_{13} & C_{14} & C_{15} & C_{16} \\ C_{21} & C_{22} & C_{23} & C_{24} & C_{25} & C_{26} \\ C_{31} & C_{32} & C_{33} & C_{34} & C_{35} & C_{36} \\ C_{41} & C_{42} & C_{43} & C_{44} & C_{45} & C_{46} \\ C_{51} & C_{52} & C_{53} & C_{54} & C_{55} & C_{56} \\ C_{61} & C_{62} & C_{63} & C_{64} & C_{65} & C_{66} \end{bmatrix} \tag{4.2}$$

where σ_{ij} and ε_{ij} are the stress and strain components, respectively, and C_{ij} are the elastic stiffness constants. The stiffness matrix is symmetric from an energy consideration. There are 21 independent constants out of the 36 constants. The above equation can also be written in the form:

$$\{\varepsilon\} = [S]\{\sigma\} \tag{4.3}$$

where [S] is the inverse matrix of [C] and is given by:

$$[S] = \begin{bmatrix} S_{11} & S_{12} & S_{13} & S_{14} & S_{15} & S_{16} \\ S_{21} & S_{22} & S_{23} & S_{24} & S_{25} & S_{26} \\ S_{31} & S_{32} & S_{33} & S_{34} & S_{35} & S_{36} \\ S_{41} & S_{42} & S_{43} & S_{44} & S_{45} & S_{46} \\ S_{51} & S_{52} & S_{53} & S_{54} & S_{55} & S_{56} \\ S_{61} & S_{62} & S_{63} & S_{64} & S_{65} & S_{66} \end{bmatrix} \tag{4.4}$$

where S_{ij} are the elastic compliances. For small deformation in the Cartesian coordinate system, the strains can be defined as:

$$\varepsilon_{ij} = \frac{1}{2}(\frac{\partial u_i}{\partial x_j} + \frac{\partial u_j}{\partial x_i}) \tag{4.5}$$

where u_i $(i=1,2,3)$ are the displacements in the directions of the three Cartesian coordinates, and x_i $(i=1,2,3)$ are the three coordinates in the Cartesian system.

For an orthotropic material, in which there are three orthogonal symmetrical planes, we have the following Hooke's law:

$$\begin{Bmatrix} \sigma_{11} \\ \sigma_{22} \\ \sigma_{33} \\ \sigma_{23} \\ \sigma_{31} \\ \sigma_{12} \end{Bmatrix} = \begin{bmatrix} C_{11} & C_{12} & C_{13} & 0 & 0 & 0 \\ C_{12} & C_{22} & C_{23} & 0 & 0 & 0 \\ C_{13} & C_{23} & C_{33} & 0 & 0 & 0 \\ 0 & 0 & 0 & C_{44} & 0 & 0 \\ 0 & 0 & 0 & 0 & C_{55} & 0 \\ 0 & 0 & 0 & 0 & 0 & C_{66} \end{bmatrix} \begin{Bmatrix} \varepsilon_{11} \\ \varepsilon_{22} \\ \varepsilon_{33} \\ 2\varepsilon_{23} \\ 2\varepsilon_{31} \\ 2\varepsilon_{12} \end{Bmatrix} \tag{4.6}$$

in which there are only nine independent elastic stiffness constants. Similarly, there are only nine independent elastic compliance constants, and the compliance matrix for a unidirectional fibre reinforced composite material is given by:

$$[S] = [C]^{-1} = \begin{bmatrix} 1/E_1 & -v_{12}/E_1 & -v_{13}/E_1 & 0 & 0 & 0 \\ -v_{12}/E_1 & 1/E_2 & -v_{23}/E_2 & 0 & 0 & 0 \\ -v_{13}/E_1 & -v_{23}/E_2 & 1/E_3 & 0 & 0 & 0 \\ 0 & 0 & 0 & 1/G_{23} & 0 & 0 \\ 0 & 0 & 0 & 0 & 1/G_{31} & 0 \\ 0 & 0 & 0 & 0 & 0 & 1/G_{12} \end{bmatrix}$$

where E_1, E_2, E_3, G_{12}, G_{23}, G_{31}, v_{12}, v_{23} and v_{31} are engineering constants.

4.2.2 Representative Volume Element and Effective Properties

A microscopically inhomogeneous composite material can be idealised as a macroscopically homogeneous continuum when the behaviour of engineering structures made of the material can be satisfactorily retained. Such idealisation can be realised over a representative sample of the composite material. Selection of the dimensions of a representative volume is imperative. The representative volume must be sufficiently large compared to the scale of the microstructure so that it contains a sufficient number of individual constituents and microstructural features. It also must be small compared to the whole structural body so that it is entirely typical of the whole composite structure on average. For structural scales larger than the representative volume element, continuum mechanics can be used to reproduce properties of the material as a whole for structural analysis and design without considering the microstructure of the material.

For a representative volumetric element subject to an imposed macroscopically homogeneous stress or displacement field and no body forces, the average stress and strain components are defined as:

$$\bar{\sigma}_{ij} = \frac{1}{V}\int_V \sigma_{ij}dV$$
$$\bar{\varepsilon}_{ij} = \frac{1}{V}\int_V \varepsilon_{ij}dV$$

(4.7)

where σ_{ij} and ε_{ij} are the true stresses and strains in the representative volume V or micro stresses or micro strains, respectively.

When a representative volume element is subject to a prescribed displacement field on its boundary surface S in the form:

$$u_i\big|_S = \varepsilon_{ij}^0 x_j$$

(4.8)

where ε_{ij}^0 are constant strains, the average strains $\bar{\varepsilon}_{ij}$ are identical to the applied constant strains, i.e., $\bar{\varepsilon}_{ij} = \varepsilon_{ij}^0$, when there exists perfect interfacial bonding.

When a representative volume element is subject to a homogeneous stress field on its boundary surface S in the following form:

$$t_i\big|_S = \sigma_{ij}^0 n_j$$

(4.9)

where σ_{ij}^0 are constant stresses and n_i ($i=1,2,3$) are components of the unit outward normal vector to the surface of the representative volume, the average stresses are identical to the applied constant stresses, i.e., $\bar{\sigma}_{ij} = \sigma_{ij}^0$. Both conditions in equations (4.8) and (4.9) are referred to as homogeneous boundary conditions, i.e., iso-strain and iso-stress boundary conditions, respectively. It is worth pointing out that the work done by the average stresses and the average strains are identical to that done by the micro stresses and micro strains.

The effective properties are defined in terms of the relations between the average stresses and average strains over a representative volume. They can be obtained by applying boundary conditions in equations (4.8) and (4.9) as follows.

Iso-strain method: imposition of a prescribed homogeneous displacement-based boundary condition in equation (4.8) to a representative volume yields the following stress-strain relation:

$$\{\bar{\sigma}\} = [\bar{C}]\{\bar{\varepsilon}\} \tag{4.10}$$

where

$$\{\bar{\sigma}\} = \{\bar{\sigma}_{11} \quad \bar{\sigma}_{22} \quad \bar{\sigma}_{33} \quad \bar{\sigma}_{23} \quad \bar{\sigma}_{31} \quad \bar{\sigma}_{12}\}^T$$

$$\{\bar{\varepsilon}\} = \{\bar{\varepsilon}_{11} \quad \bar{\varepsilon}_{22} \quad \bar{\varepsilon}_{33} \quad \bar{\varepsilon}_{23} \quad \bar{\varepsilon}_{31} \quad \bar{\varepsilon}_{12}\}^T$$

$$[\bar{C}] = \begin{bmatrix} \bar{C}_{11} & \bar{C}_{12} & \bar{C}_{13} & \bar{C}_{14} & \bar{C}_{15} & \bar{C}_{16} \\ \bar{C}_{21} & \bar{C}_{22} & \bar{C}_{23} & \bar{C}_{24} & \bar{C}_{25} & \bar{C}_{26} \\ \bar{C}_{31} & \bar{C}_{32} & \bar{C}_{33} & \bar{C}_{34} & \bar{C}_{35} & \bar{C}_{36} \\ \bar{C}_{41} & \bar{C}_{42} & \bar{C}_{43} & \bar{C}_{44} & \bar{C}_{45} & \bar{C}_{46} \\ \bar{C}_{51} & \bar{C}_{52} & \bar{C}_{53} & \bar{C}_{54} & \bar{C}_{55} & \bar{C}_{56} \\ \bar{C}_{61} & \bar{C}_{62} & \bar{C}_{63} & \bar{C}_{64} & \bar{C}_{65} & \bar{C}_{66} \end{bmatrix} \tag{4.11}$$

where $\bar{\sigma}_{ij}$ and $\bar{\varepsilon}_{ij}$ are the average stress and average strain components, respectively, and $\bar{C}_{ij}$ are the effective elastic stiffness constants.

Iso-stress method: application of homogeneous boundary conditions in equation (4.9) to a representative volume leads to the following stress-strain relation:

$$\{\bar{\varepsilon}\} = [\bar{S}]\{\bar{\sigma}\} \tag{4.12}$$

where

$$[\bar{S}] = \begin{bmatrix} \bar{S}_{11} & \bar{S}_{12} & \bar{S}_{13} & \bar{S}_{14} & \bar{S}_{15} & \bar{S}_{16} \\ \bar{S}_{21} & \bar{S}_{22} & \bar{S}_{23} & \bar{S}_{24} & \bar{S}_{25} & \bar{S}_{26} \\ \bar{S}_{31} & \bar{S}_{32} & \bar{S}_{33} & \bar{S}_{34} & \bar{S}_{35} & \bar{S}_{36} \\ \bar{S}_{41} & \bar{S}_{42} & \bar{S}_{43} & \bar{S}_{44} & \bar{S}_{45} & \bar{S}_{46} \\ \bar{S}_{51} & \bar{S}_{52} & \bar{S}_{53} & \bar{S}_{54} & \bar{S}_{55} & \bar{S}_{56} \\ \bar{S}_{61} & \bar{S}_{62} & \bar{S}_{63} & \bar{S}_{64} & \bar{S}_{65} & \bar{S}_{66} \end{bmatrix} \tag{4.13}$$

where $\bar{S}_{ij}$ are the effective elastic compliances.

For a representative volume element, with well-defined dimensions, of a composite material, the effective properties produced by employing the homogeneous boundary conditions given in equations (4.8) and (4.9) are expected to be identical, i.e., the effective compliance matrix, $[\overline{S}]$, is the inverse matrix of the effective elastic stiffness matrix $[\overline{C}]$. When applying the boundary conditions in equations (4.8) or (4.9) to a representative volume, solutions for the true stresses or true strains can be obtained either analytically or numerically. In the analytical approaches, various assumptions are introduced to simplify the solutions, which in turn yield simple and closed-form expressions for the effective properties of the representative volume. However, the selected assumptions may not allow consideration of certain characteristics and their corresponding parameters. In the numerical analysis approach, finite element methods may be used and lesser number of assumptions is required in the analysis, which allows consideration of more characteristics of a representative volume element. With the advance of computing techniques, numerical simulation can be achieved at multi-scales, which allows modelling of more features. However, numerical analysis approaches can be expensive and require more microstructural information of a representative volume.

4.2.3 Rules of Mixtures and Mori-Tanaka Theory

As an illustrative example, consider a two-phase composite consisting of an elastic matrix reinforced by randomly dispersed spherical elastic inclusions. The average stress and strain are given by:

$$\{\overline{\sigma}\} = c_1\{\overline{\sigma}^{(1)}\} + c_2\{\overline{\sigma}^{(2)}\}$$
$$\{\overline{\varepsilon}\} = c_1\{\overline{\varepsilon}^{(1)}\} + c_2\{\overline{\varepsilon}^{(2)}\} \tag{4.14}$$

where c_1, c_2 are the volume fraction of each phase with $c_1+c_2=1$, $\overline{\sigma}^{(i)}$ and $\overline{\varepsilon}^{(i)}$ ($i=1,2$) are the average stress and strain vectors in phase 1 and 2 respectively.

Using the relations between stresses and strains at any point in the phase as given in equation (4.10) and (4.11), the above equations can be written as:

$$\{\overline{\sigma}\} = c_1[C^{(1)}]\{\overline{\varepsilon}^{(1)}\} + c_2[C^{(2)}]\{\overline{\varepsilon}^{(2)}\}$$
$$\{\overline{\varepsilon}\} = c_1[S^{(1)}]\{\overline{\sigma}^{(1)}\} + c_2[S^{(2)}]\{\overline{\sigma}^{(2)}\} \tag{4.15}$$

The average strains and stresses in each phase are uniquely dependent on the average strains and stresses in representative volume element, namely,

$$\{\overline{\varepsilon}^{(i)}\} = [A_i]\{\overline{\varepsilon}\} \qquad \{\overline{\sigma}^{(i)}\} = [B_i]\{\overline{\sigma}\} \qquad (i=1,2) \tag{4.16}$$

where A_i and B_i ($i=1,2$) are referred to as concentration matrices, and $c_1[A_1]+c_2[A_2]=[I]$ and $c_1[B_1]+c_2[B_2]=[I]$.

Substituting equations (4.16) into (4.15) yields the following expressions for the effective stiffness and compliance matrices of the composite material:

$$[\overline{C}] = c_1[C^{(1)}][A_1] + c_2[C^{(2)}][A_2] \tag{4.17a}$$

$$[\bar{S}] = c_1[S^{(1)}][B_1] + c_2[S^{(2)}][B_2] \tag{4.17b}$$

Noting the definition of concentration matrices in equation (4.16), the following relationships hold $c_1[A_1] = [I] - c_2[A_2]$ and $c_1[B_1] = [I] - c_2[B_2]$. Thus the above equations can be rewritten as:

$$[\bar{C}] = [C^{(1)}] + c_2\{[C^{(2)}] - [C^{(1)}]\}[A_2]$$
$$[\bar{S}] = [S^{(1)}] + c_2\{[S^{(2)}] - [S^{(1)}]\}[B_2] \tag{4.18}$$

By assuming that the strain is uniform throughout the composite, which means that $[A_1]=[A_2]=[I]$, the following simplest equation can be obtained:

$$[\bar{C}] = c_1[C^{(1)}] + c_2[C^{(2)}] \tag{4.19}$$

Similarly, by assuming that stress is uniform throughout the composites, namely $[B_1]=[B_2]=[I]$, we have the following:

$$[\bar{S}] = c_1[S^{(1)}] + c_2[S^{(2)}] \tag{4.20}$$

Equation (4.19) and (4.20) are the Voigt and Reuss approximations, which provide upper and lower bounds as proved by Hill (Aboudi 1991).

    Determination of the concentration matrices in different phases is one of the most important steps in evaluating the effective overall properties of a composite material. Mori and Tanaka (1973) presented a method for calculating average internal stress in a matrix of materials containing misfitting inclusions by using eigenstrains. In the Mori-Tanaka method, the average strain in the interacting inclusions is approximated by that of a single inclusion in an infinite matrix subject to the uniform average matrix strain (Aboudi, 1991), which leads to the following relation:

$$\{\bar{\varepsilon}^{(2)}\} = [T_S]\{\bar{\varepsilon}^{(1)}\}$$

where superscript 2 indicates the inclusion and superscript 1 corresponds to the matrix, $[T_S]$ is determined from the solution of a single particle imbedded in an infinite matrix subject to homogeneous displacement boundary conditions defined by the average matrix strain $\{\bar{\varepsilon}^{(1)}\}$. Substituting the above equation into (4.15) yields a definition of $\{\bar{\varepsilon}^{(1)}\}$ in terms of the overall average strains $\{\bar{\varepsilon}\}$, and in conjunction with equation (4.15) leads to the determination of $[A_2]$ as follows (Aboudi 1991):

$$[A_2] = [T_S](c_1[I] + c_2[T_S])^{-1}$$

Substituting into equation (4.18) yields the overall stiffness matrix. The overall compliance matrix can also be obtained similarly.

4.2.4 Unit Cell Models for Textile Composites

As described in Chapter 2, textile composites, including two-dimensional woven and braided composites, are manufactured with advanced machinery following specifically designed parameters. Such manufacturing processes result in textile composites possessing geometric periodic patterns, i.e., there exists a piece of minimum sized sample of composite which can be copied with translational increments only repetitively to map out the whole composite structures. For a fibre reinforced textile composite material, the minimum sized periodic sample is chosen as the unit cell of the material because it is small and also contains all individual constituents and microstructural features. Unit cell approach has been widely used in almost all available micromechanics models developed for fibre reinforced textile composites (Chou and Ko 1989; Tan et al., 1997a; Mouritz et al., 1999; Tan, 1999).

Prediction of the effective properties for a unit cell to a fibre reinforced composite material proved and remains to be a great challenge. Presentation of all micromechanics models available is a daunting task. To present some of the basic concepts and ideas, we choose to divide all models into two categories, i.e., analytical or semi-analytical approach and numerical approach based on finite element methods (FEM). In analytical models, simple formulas may be obtained for the effective properties based primarily on a large number of assumptions. In the numerical based models, effective properties can be evaluated numerically only by taking into account more detailed features of the microstructure, such as fibre tow architectures, using the finite element method.

4.3 UNIT CELL MODELS FOR 2D WOVEN COMPOSITES

Two-dimensional woven composites are produced on a loom that interlaces two sets of fibre yarns at right angle to each other. The lengthwise yarns are referred to as warps, while the yarns perpendicular to the warps are called fills or wefts. Each yarn is a bundle, and its size is related to the number of fibres in the yarn, the diameter of the fibres, and the packing density of fibres. Figure 4.1 depicts schematically the top views of some commonly used 2D woven composites and the cross-sectional views of the weaves. The various types of woven composites can be readily identified by the patterns of repeats in both warp and weft directions, defined by two geometrical quantities n_g^w and n_g^f. The number of n_g^w means that a weft (fill) yarn is interspersed with every n_g^w-th warp yarn, while the number of n_g^f indicates that a warp yarn is interlaced with every n_g^f-th weft (fill) yarn. For all weaves in Figure 4.1, the two geometrical quantities are identical, i.e. $n_g^w = n_g^f = n_g$. The plain weave has a tighter interlacing, while the twill and satin weaves have a looser interlacing. The interlacing of the yarn causes the yarn undulation or yarn crimp.

There has been extensive research on the prediction of effective properties for 2D fibre reinforced woven composite materials. It is not the intent of this book to include all published models; instead, we chose to present some of the widely known models by classifying them into one-, two- and three-dimensional models as well as the applications of finite element method.

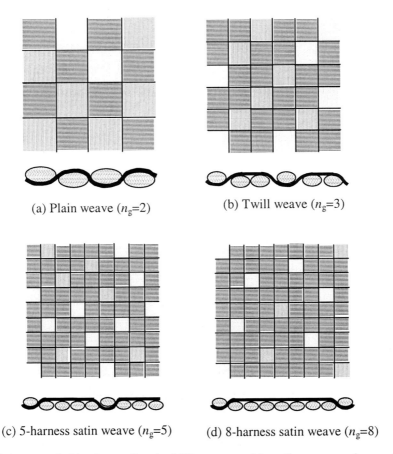

(a) Plain weave (n_g=2)   (b) Twill weave (n_g=3)

(c) 5-harness satin weave (n_g=5)   (d) 8-harness satin weave (n_g=8)

Figure 4.1 Top and side views of typical 2D weaves with weft yarns running vertically and warp yarn running horizontally and their unit cells

4.3.1 One-Dimensional (1D) Models

In early 1980's, Ishikawa and Chou (1982a,b; 1983a,b,c) developed three basic analytical models, i.e., mosaic model, fibre undulation model and bridging model, for predicting the thermo-elastic behaviour of woven composites.

In the mosaic model, a woven composite is idealised as an assemblage of asymmetrical cross-ply laminates. Figure 4.2 illustrates the side view of the mosaic model for a repeating unit for an 8-harness satin woven composite. The cross-ply laminate is then modelled based on the classical laminated plate theory neglecting the shear deformation in the thickness direction (Jones, 1975). The constitutive equations are given by

$$\begin{Bmatrix} N \\ M \end{Bmatrix} = \begin{bmatrix} A_{ij} & B_{ij} \\ B_{ij} & D_{ij} \end{bmatrix} \begin{Bmatrix} \varepsilon \\ \kappa \end{Bmatrix} \tag{4.21}$$

where N and M are the membrane stress resultant and bending moment vectors, ε and κ are the vectors of in-plane strains and changes of curvature on the middle plane of the laminate, and A_{ij}, B_{ij} and D_{ij} (i,j=1,2,6) are the in-plane stretching, bending/stretching coupling, and bending stiffness matrices, which can be calculated using:

$$\left(A_{ij} \quad B_{ij} \quad D_{ij}\right) = \int_{-\frac{1}{2}}^{\frac{1}{2}} \overline{Q}_{ij}\left(1 \quad z \quad z^2\right)dz \tag{4.22}$$

where $\overline{Q}_{ij}$ are the elastic constants of a lamina, which is a function of fibre orientation, (see Jones, 1975).

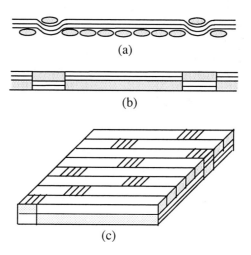

(a)

(b)

(c)

Figure 4.2 Mosaic Model for woven composites (Ishikawa and Chou, 1982a)

As a first order approximation, an iso-strain field is assumed in the middle plane. The effective stiffness constants for the unit cell of a 2D woven composite idealised as an asymmetrical cross-ply laminate can be expressed as:

$$\overline{A}_{ij} = \frac{1}{A_{uc}} \iint_{A_{uc}} A_{ij}\,dA, \quad \overline{B}_{ij} = \frac{1}{A_{uc}} \iint_{A_{uc}} B_{ij}\,dA, \quad \overline{D}_{ij} = \frac{1}{A_{uc}} \iint_{A_{uc}} D_{ij}\,dA \tag{4.23}$$

Where the bar on top of a variable indicates an average of that variable over the area of the unit cell A_{uc}. When these stiffness constants are inverted, lower bounds of the elastic compliance constants can also be obtained. For non-hybrid weaves, the averages can be simplified as:

$$\overline{A}_{ij} = A_{ij}^*, \quad \overline{B}_{ij} = (1 - \frac{2}{n_g})B_{ij}^*, \quad \overline{D}_{ij} = D_{ij}^* \qquad (4.24)$$

where A_{ij}^*, B_{ij}^* and D_{ij}^* are the stretching, stretching-bending coupling and bending stiffness matrices of a two-ply cross-ply asymmetrical laminate.

The constitutive equations (4.21) can also be written in an inverted form as follows:

$$\begin{Bmatrix} \varepsilon \\ \kappa \end{Bmatrix} = \begin{bmatrix} a_{ij} & b_{ij} \\ b_{ij} & d_{ij} \end{bmatrix} \begin{Bmatrix} N \\ M \end{Bmatrix} \qquad (4.25)$$

Application of iso-stress field to the middle plane yields the following equations for the effective compliance constants:

$$\overline{a}_{ij} = \frac{1}{A_{uc}} \iint_{A_{uc}} a_{ij} dA, \quad \overline{b}_{ij} = \frac{1}{A_{uc}} \iint_{A_{uc}} b_{ij} dA, \quad \overline{d}_{ij} = \frac{1}{A_{uc}} \iint_{A_{uc}} d_{ij} dA \qquad (4.26)$$

The above equations provide upper bounds of the compliance constants and lower bounds of the stiffness constants when inverted. For non-hybrid weaves, the averages can be simplified as:

$$\overline{a}_{ij} = a_{ij}^*, \quad \overline{b}_{ij} = (1 - \frac{2}{n_g})b_{ij}^*, \quad \overline{d}_{ij} = d_{ij}^* \qquad (4.27)$$

where a_{ij}^*, b_{ij}^* and d_{ij}^* are the stretching, stretching-bending coupling and bending compliance matrices of a two-ply cross-ply asymmetrical laminate.

Mosaic model provides upper and lower bounds for the effective stiffness and compliance constants for a unit cell of woven composite. However, fibre continuity and non-uniform stresses and strains in the interlaced region are not considered although a good agreement between predictions and experimental results was reported. It is clear that fibre continuity and undulation are not taken into account in the idealisation process. Consequently, a one-dimensional crimp model named as "fibre undulation model" was proposed that takes into account the fibre continuity and undulation omitted in the "mosaic model".

Figure 4.3 depicts the concept of the fibre undulation model. In this model, it is assumed that the geometry of fibre undulation in the weft yarn can be expressed in the form of the following sinusoidal function within the length of a_u:

$$h_1(x) = \frac{h_t}{4}\left[1 - \sin\left\{(x - \frac{a}{2})\frac{\pi}{a_u}\right\}\right] \qquad (4.28)$$

and the sectional shape of the warp yarn is assumed to take the following form:

$$h_2(x) = \begin{cases} \dfrac{h_t}{4}\left[1-\sin\left\{(x-\dfrac{a}{2})\dfrac{\pi}{a_u}\right\}\right] & \text{when } a_0 \leq x \leq a/2 \\[3mm] \dfrac{h_t}{4}\left[-1-\sin\left\{(x-\dfrac{a}{2})\dfrac{\pi}{a_u}\right\}\right] & \text{when } a/2 \leq x \leq a_2 \end{cases}$$ (4.29)

where $a_0 = (a-a_u)/2$ and $a_2 = (a+a_u)/2$. Clearly, both functions are independent of the y-axis, which indicates that the fibre undulation in the warp direction is neglected.

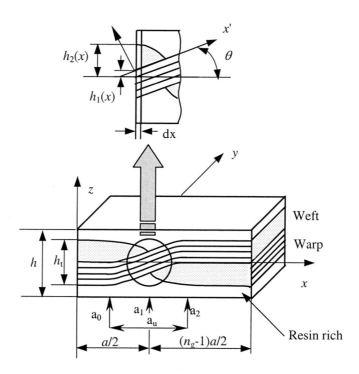

Figure 4.3 Fibre undulation model (Ishikawa and Chou, 1982b)

The unit cell in Figure 4.3 consists of two straight cross-ply regions and one undulated cross-ply region. For the undulated portion, the angle θ between the local fibre orientation and the global coordinate is a function of x only and can be determined from the function $h_1(x)$. The elastic stiffness constants $\overline{Q}_{ij}^F$ for the undulated weft yarn can be expressed in terms of the elastic stiffness constants Q_{ij}^F and the angle θ, and these are defined as follows by Ishikawa and Chou (1982b):

$$Q_{ij}^F(\theta) = \begin{bmatrix} E_x^F(\theta)/D_v & E_y^F v_{yx}^F(\theta)/D_v & 0 \\ E_y^F v_{yx}^F(\theta)/D_v & E_y^F/D_v & 0 \\ 0 & 0 & G_{xy}^F(\theta) \end{bmatrix} \tag{4.30}$$

where $i,j=1,2,6$ and

$$E_x^F(\theta) = 1/[l_\theta^4/E_x^F + (1/G_{xz}^F - 2v_{yx}^F/E_x^F)l_\theta^2 m_\theta^2 + m_\theta^4/E_z^F]$$
$$v_{yx}^F(\theta) = v_{zx}^F l_\theta^2 + v_{yz}^F m_\theta^2$$
$$G_{xy}^F(\theta) = G_{xy}^F l_\theta^2 + G_{yz}^F m_\theta^2$$
$$E_y^F(\theta) = E_y^F = E_z^F$$
$$D_v = 1 - v_{yx}^F(\theta)^2 E_y^F/E_x^F(\theta) \tag{4.31}$$

The angle θ can be expressed in terms of x as follows:

$$\theta(x) = \arctan\left(\frac{dh_1(x)}{dx}\right) \tag{4.32}$$

The following formulas were obtained by Ishikawa and Chou (1982b):

$$\begin{aligned} A_{ij}(x) &= Q_{ij}^W[h_2(x) - h_1(x)] + Q_{ij}^F(x)h_t/2 \\ &\quad + Q_{ij}^M[h_1(x) - h_2(x) + h - h_t/2] \\ B_{ij}(x) &= \tfrac{1}{4}Q_{ij}^W[h_2(x) - h_1(x)]h_t + \tfrac{1}{2}Q_{ij}^F(x)[h_1(x) - h_t/4]h_t \\ D_{ij}(x) &= \tfrac{1}{3}Q_{ij}^M\{[h_1(x) - h_t/2]^3 - h_2^3(x) + h^3/4\} \\ &\quad + \tfrac{1}{3}Q_{ij}^F(x)[h_t^3/8 - 3h_t^2 h_1(x)/4 + 3h_t h_1^2(x)/2] \\ &\quad + \tfrac{1}{3}Q_{ij}^W[h_2^3(x) - h_1^3(x)] \end{aligned} \tag{4.33}$$

where the superscripts F, W and M represent the weft, warp and matrix, respectively. By inverting the matrices defined by equation (4.33), the corresponding compliance matrices, $a_{ij}^{*u}(x)$, $b_{ij}^{*u}(x)$ and $d_{ij}^{*u}(x)$ for the undulated portion can be obtained. Solutions based on the assumption of uniform stress for the infinitesimal pieces in the straight and crimped regions, as shown in Figure 4.3, were assembled, and the average compliance properties can be obtained as follows:

$$\bar{a}_{ij} = (1 - \frac{2a_u}{n_g a})a_{ij}^* + \frac{2}{n_g a}\int_{a_0}^{a_2} a_{ij}^{*u}(x)dx$$

$$\bar{b}_{ij} = (1 - \frac{2}{n_g})b_{ij}^* + \frac{2}{n_g a}\int_{a_0}^{a_2} b_{ij}^{*u}(x)dx$$

$$\overline{d}_{ij} = (1 - \frac{2a_u}{n_g a})d_{ij}^* + \frac{2}{n_g a} \int\limits_{a_0}^{a_2} d_{ij}^{*u}(x)dx \tag{4.34}$$

where a_{ij}^*, b_{ij}^* and d_{ij}^* are the stretching, stretching-bending coupling and bending compliance matrices of a two-ply cross-ply asymmetrical laminate, i.e., the straight portion. a_{ij}^{*u}, b_{ij}^{*u} and d_{ij}^{*u} are those for the undulated portion. This model is an extension of the series model and is applicable to weaves with low n_g values, i.e., $n_g=2$.

Both the mosaic model and fibre undulation model are useful for understanding the basic aspects of the mechanical properties for woven fabrics. For example, it was found that the relationship between the in-plane stiffness C_{11} and $1/n_g$ by applying the well-known "mosaic model" and "fibre undulation model" shows that the reduction in C_{11} due to fibre undulation is most severe in plain woven ($n_g=2$) as compared to cross-ply laminates ($1/n_g=0$, ie., straight yarn).

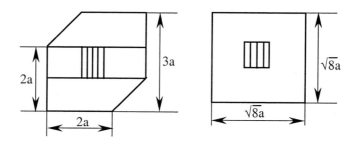

Figure 4.4 Unit cell for 8 harness satin weave (Ishikawa and Chou, 1982b)

Ishikawa and Chou (1982b) also proposed a bridging model, in which the interactions between an undulated region and its surrounding regions with straight threads were considered. For 8 harness satin weaves, the unit cell can be chosen as a hexagonal shape of the repeating unit, as shown in Figure 4.4, which can be transformed into a square shape of the same area for simplicity of calculations. Figure 4.5 illustrates the concept of the bridging model, which decomposes the square unit cell into five subregions for determining the effective properties. The four regions denoted by I, II, IV and V consist of straight threads, and hence are regarded as pieces of cross-ply laminates. Region III has an interlaced structure where only the weft yarn is assumed to be undulated, since the effect of the undulation and continuity in the warp yarns is expected to be small when a load is applied load in weft direction. When it is assumed that region II, III and IV are under the same average mid-plane strain and curvature, i.e., iso-strain condition, the average stiffness constants for the assembled region II, III and IV can be obtained. The corresponding average compliance constants can then be determined by inverting the average stiffness constants. By further assuming that region I, V and the assembled region II, III and IV are under the same average mid-plane stress resultants, i.e., iso-stress condition, the average compliance constants for the whole unit cell can be determined. It is clear a combination of iso-strain and iso-stress conditions are used in the bridging model. The bridging model is considered to be applicable to satin weave with $n_g \geq 4$.

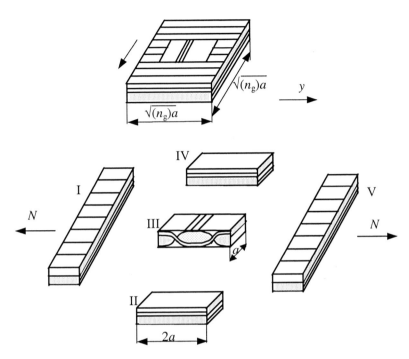

Figure 4.5 Bridging model (Ishikawa and Chou, 1982b)

Ishikawa and Chou (1982b) applied the "bridging model" to investigate the linear elastic properties of woven fabrics and non-linear behaviour due to the initial failure of the fabrics. It was reported that the elastic stiffness and knee stress in satin weave composites were higher than those in plain weave composites due to the presence of bridging regions in the weaving pattern. The fibre undulation model and bridging model were applied to analyse the non-linear elastic behaviour of fabric composites (Ishikawa and Chou, 1983b), coupling with the non-linear constitutive relation developed by Hahn and Tsai (1973). However, only the undulation and continuity of yarns along the loading direction were considered, and the yarn undulation in the transverse direction and its actual cross-sectional geometry were neglected.

Ishikawa and his colleagues (1985) conducted experiments to verify the theoretical predictions obtained in their previous work. In these experimental tests, the maximum strain level of 500×10^{-6} was chosen. The materials used were plain weave and 8-harness satin fabric composites of carbon/epoxy. It was found that for plain weave composites, the elastic moduli increases with the laminate ply number but levels out at about 8-ply thickness. The ratio of ply thickness to thread width (i.e., h/a) is also a very important parameter, which strongly affects the elastic moduli of plain weave composites. In-plane shear modulus decreases almost linearly with the fibre volume fraction which decreases with n_g.

4.3.2 Two-Dimensional (2D) Models

The fibre undulation model (Ishikawa and Chou, 1982b) considered fibre continuity and fibre undulation in one direction only, and is thus deemed as a 1D model. 2D models should take into account fibre undulation and continuity in both the warp and weft directions. In 1992, Naik and colleagues (Naik and Shembekar, 1992a,b; Shembekar and Naik, 1992 and Naik and Ganesh, 1992) extended the fibre undulation model and developed 2D models, which includes the fibre undulation and continuity in both warp and weft directions, the possible presence of a gap between adjacent yarns, and the actual cross-sectional geometry of fibre yarns. To present the fundamental concepts of 2D models, let us consider a representative cell of a plain wave lamina, as shown in Figure 4.6, for a plain weave shown in Figure 4.1.

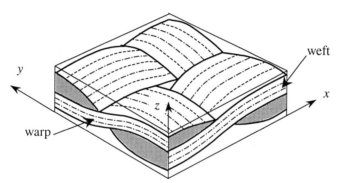

Figure 4.6 The repetitive unit cell of plain weave lamina

The unit cell consists of two fibre yarns, warp and weft, and pure matrix regions. It is desirable to obtain accurate geometrical descriptions of individual fibres or even the warp and weft yarns themselves in the space. Due to the nature of manufacturing process, the geometry of fibres or fibre yarns inevitably vary from one cell to another, thus assumptions must be introduced based on experimental observations to simplify the geometrical visualisation problem of fibres or fibre yarns. One assumption is to assume that the repetitive unit cell possesses two planes of symmetry in the interlacing region. By virtue of the symmetry, Naik and Ganesh (1992) considered only one quarter of the repetitive unit cell as shown in Figure 4.7(a) and proposed two models based on the classical laminate theory, one is referred to as slice array model and the other element array model.

In the slice array model, the unit cell is discretised into slices, for example three slices as shown in Figure 4.7(b), along the loading direction (y direction in this case). Each slice is then transformed into a four-layered laminate, i.e., an asymmetrical cross ply sandwiched between two pure matrix layers as shown in Figure 4.7(c). The effective elastic constants of the plain weave lamina are evaluated from the properties of each individual laminate slice, which are determined by considering the presence of fibre undulation.

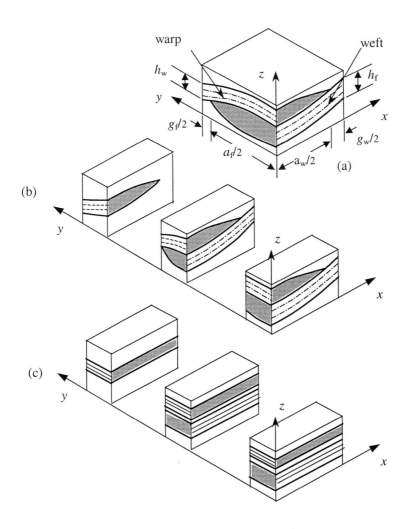

Figure 4.7 Slice array model for unit cell of plain weave: (a) Unit cell of plain weave lamina, (b) actual slices along y coordinate of the unit cell, (c) idealised slices for the actual slices (Naik and Ganesh, 1992)

It is assumed that the yarn cross-sectional shape is uniform and the fibres in the yarn are undulated only in the longitudinal direction. To describe the uniform cross-sectional shape, as shown in Figure 4.7(a), four parameters are used to define the length $0.5(a_w+g_w)$ and width $0.5(a_f+g_f)$ of the unit cell, and three parameters are used to define the thickness of the unit cell $0.5h_m+h_f+h_w+0.5h_m$. Based on the uniform cross-sectional shape assumption, it is clear that the height of the weft is a function of only y and the height of the warp yarn is a function of only x. Similar to the functions defined in both y-z and z-x planes by Naik and Ganesh (1992), the following three different sinusoidal functions may be utilised to define the shape and the yarn undulation:

$$z_1(x, y) = \frac{h_f + h_w + h_m}{2} + \frac{h_f}{2}\cos\left(\frac{\pi y}{a_f + g_f}\right) + \frac{h_w}{2}\cos\left(\frac{\pi x}{a_{xt}}\right)$$

$$z_2(x, y) = \frac{h_f + h_w + h_m}{2} + \frac{h_f}{2}\cos\left(\frac{\pi y}{a_f + g_f}\right) - \frac{h_w}{2}\cos\left(\frac{\pi x}{a_w + g_w}\right)$$

$$z_3(x, y) = \frac{h_f + h_w + h_m}{2} - \frac{h_f}{2}\cos\left(\frac{\pi y}{a_{yt}}\right) - \frac{h_w}{2}\cos\left(\frac{\pi x}{a_w + g_w}\right) \qquad (4.35)$$

where

$$a_{xt} = \frac{a_w(a_w + g_w)}{a_w + 2g_w} \qquad\qquad a_{yt} = \frac{a_f(a_f + g_f)}{a_f + 2g_f} \qquad (4.36)$$

In equation (4.35), $z_1(x,y)$ defines the upper surface of the warp yarn, $z_2(x,y)$ defines the lower surface of the warp yarn and also the upper surface of the weft yarn, and $z_3(x,y)$ defines the lower surface of the weft yarn. The ranges of x and y are [0, 0.5a_w; 0, 0.5(a_f+g_f)] for the warp yarn and [0, 0.5(a_w+g_w); 0, 0.5a_f] for the weft yarn, respectively. The heights of the warp and weft yarns are given by

$$h_{wp}(x) = z_1(x, y) - z_2(x, y) = \frac{h_w}{2}\cos\left(\frac{\pi x}{a_{xt}}\right) + \frac{h_w}{2}\cos\left(\frac{\pi x}{a_w + g_w}\right)$$

$$h_{wf}(y) = z_2(x, y) - z_3(x, y) = \frac{h_f}{2}\cos\left(\frac{\pi y}{a_{yt}}\right) + \frac{h_f}{2}\cos\left(\frac{\pi y}{a_f + g_f}\right) \qquad (4.37)$$

and equation (4.36) was obtained by setting the height of both yarns to be zero. The undulation function is defined as the trajectory of the centre of an individual yarn. The undulation functions for both warp and weft yarns are given by:

$$z_{wp}(y) = 0.5(z_1(0, y) + z_2(0, y)) = \frac{h_f + h_w + h_m}{2} + \frac{h_f}{2}\cos\left(\frac{\pi y}{a_f + g_f}\right)$$

$$z_{wf}(x) = 0.5(z_1(x,0) + z_2(x,0)) = \frac{h_f + h_w + h_m}{2} - \frac{h_w}{2}\cos\left(\frac{\pi x}{a_w + g_w}\right) \qquad (4.38)$$

Differentiating the two undulation functions, we can readily obtain the expressions for the fibre orientation angles, $\theta_{wp}(y)$ and $\theta_{wf}(x)$, of the warp and weft yarns in relation to the global x and y coordinates. With the off-axis fibre orientation angle known for each yarn, the reduced compliance constants, $S_{ij}(\theta)$, of the undulated yarns along the global axes can be determined using transformed equations similar to those given in (4.30) and (4.31).

As proposed by Naik and Ganesh (1992), with reference to Figure 4.7(b) and (c), for the warp yarn in one slice the off-axis angle at the midpoint of that slice is used to calculate the off-axis compliance constants for the idealised warp layer, while for the weft yarn in one slice the off-axis effective compliance constants for the idealised weft layer are set to equal to the average values of $S_{ij}(\theta)$ over the integration interval from 0 to the maximum values of $\theta_{wf}(x)$. With the compliance constants known for each layer in one idealised slice in Figure 4.7(c), the effective properties for that slice can be determined using the classical laminate theory. Similar to the case of one-dimensional fibre undulation model, the overall effective properties for the unit cell shown in Figure 4.7(a) can be determined by applying iso-stress conditions to all slices in y direction.

Evidently, determination of the off-axis compliance constants in the weft yarn in terms of average values of $S_{ij}(\theta)$ introduces approximations. Another model, proposed by Naik and Ganesh (1992) and referred to as the element array model, was to enhance the approximation in the weft direction. In the element array model, the slices of the unit cell shown in Figure 4.7(b) were further divided into elements along the x direction prior to idealisation. This can be better illustrated in Figure 4.8, in which the unit cell is discretised along both warp and weft directions into elements. For each element the off-axis angles at the centre of the element are chosen to determine the reduced properties of the idealised layers, which can be further used to calculate the properties of that element using the classical laminate theory. The overall effective properties of the unit cell can be obtained by assembling all elements in two combinations, i.e., series-parallel combination and parallel-series combination. In the series-parallel combination, elements are assembled in series into slices first along the loading direction under iso-stress condition and then the slices are considered in parallel under iso-strain condition. In the parallel-series combination, elements are grouped in parallel into slices first across the loading direction under an iso-strain condition and then the slices are considered in series under an iso-stress condition. It is expected that the parallel-series combination predict a higher value of stiffness compared to the series-parallel combination.

Naik and his colleagues have performed an extensive research both numerically and experimentally to verify the slices array model and element array model (Naik, 1994). Shembekar and Naik (1992) also investigated the effect of fibre undulation shifts between individual weave lamina in a laminated plate. It is beyond the scope and limit of this chapter. Readers who require further details on the models themselves and experimental verification are referred to the book by Naik (1994).

4.3.3 Three-Dimensional (3D) Models

Both 1D and 2D models discussed above were developed based on the classical laminate theory. Although accounting for yarn undulation, yarn shape and spacing, these models predict the in-plane elastic properties only. 3D models have been developed to evaluate the out-of-plane elastic properties in addition to the in-plane properties. It is not possible to present all models. In the following we choose to present the models proposed by Hahn and Pandey (1994) and Vandeurzen et al. (1996a, 1996b, 1998) for the case of plain weave composites.

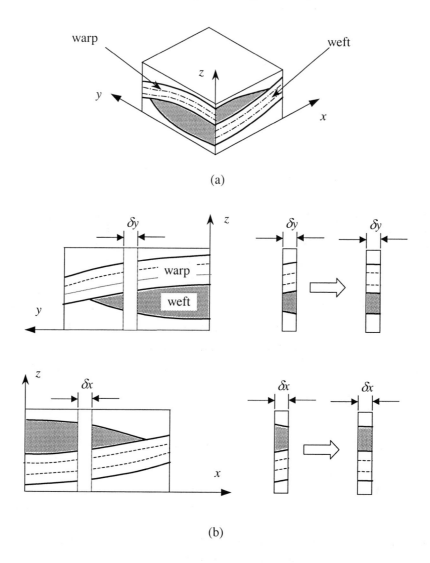

(a)

(b)

Figure 4.8 Element array model (a) unit cell, (b) division and idealisation in z-x and z-y planes (Naik and Ganesh, 1992)

Consider the unit cell shown in Figure 4.6, we assume that the undulation and cross-sectional shape of each individual yarn is known. Both yarns are treated as transversely isotropic unidirectional composites, for which the engineering constants can be measured or evaluated from the local fibre volume fraction and fibre and matrix properties. There exist a number of closed-form approximations for evaluating the properties of unidirectional composites (Hill 1965; Hashin 1979; Christensen 1990; Aboudi, 1991;). For example, Chamis (1984) proposed the following equations for the engineering constants:

$$E_{11} = V_f E_{f1} + (1 - V_f) E_m$$

$$E_{22} = E_{33} = \frac{E_m}{[1 - \sqrt{V_f}(1 - E_m/E_{f2})]}$$

$$G_{12} = G_{13} = \frac{G_m}{[1 - \sqrt{V_f}(1 - G_m/G_{f12})]} \qquad (4.39)$$

$$G_{23} = \frac{G_m}{[1 - \sqrt{V_f}(1 - G_m/G_{f23})]}$$

$$V_{12} = V_{13} = V_f V_{f12} + (1 - V_f) V_m$$

where E_m, G_m and v_m are the matrix elastic properties, E_{f1}, E_{f2}, G_{f12}, G_{f23} and v_{f12} are the fibre elastic properties, and V_f represents the fibre volume fraction of the yarn. In terms of the engineering constants, the corresponding stiffness matrix $[C]$ and the compliance matrix $[S]$ can be determined following the standard procedure (Christensen, 1979) in the yarn-related local coordinate system. From the yarn undulation function, the off-axis angle can be determined and then used to compute the off-axis elastic stiffness matrix, which can be a function of x or y. With reference to Figure 4.6, the off-axis stiffness and compliance matrix for the warp are $[C^w(y)]$ and $[S^w(y)]$, and those for the weft are $[C^f(x)]$ and $[S^f(x)]$, by employing the transformation matrix $[T]$ defined between the yarn-based local coordinate system and the global coordinate system for the unit cell.

In the 3D model proposed by Hahn and Pandey (1994) the representative volume element as shown in Figure 4.6 is considered but without modelling the upper and lower layer of pure matrix. Yarn undulations and geometry are described by using the sinusoidal functions similar to those given in equations (4.35), (4.37) and (4.38) with $h_m=0$, h_w and $h_f=0$ being exchanged, L_f replacing a_{xt} and a_w+g_w, and a_{yt} and L_w replacing a_f+g_f. Average stresses and strains are defined as those in equation (4.7). To simplify the analysis, it is assumed that the strains are uniform throughout the unit cell when it is subject to homogenous displacement boundary conditions similar to the definition in equation (4.8). This is a key assumption in this model as it introduces the approximation. Under this iso-strain assumption, the effective elastic properties $[\overline{C}]$ as defined in equation (4.10) can be determined by:

$$[\overline{C}] = \frac{1}{V}\left[\int_{V_w} [C^w(y)]dV + \int_{V_f} [C^f(x)]dV + \int_{V_m} [C^m]dV \right] \qquad (4.40)$$

where the subscripts and superscripts w, f and m represent the warp, weft and pure matrix, respectively, V is the total volume of the unit cell. Closed form expressions for the effective elastic constant matrix $[\overline{C}]$ were given by Hahn and Pandey (1994). The iso-strain assumption offers a significant simplification in evaluating effective elastic stiffness matrix $[\overline{C}]$, but it also at the same time creates an opportunity for future research to enhance the predicted results by removing the iso-strain assumption.

Replacing the iso-strain assumption with an iso-stress assumption throughout the unit cell, we can find the effective compliance matrix as follows:

$$[\bar{S}] = \frac{1}{V}\left[\int_{V_w}[S^w(y)]dV + \int_{V_f}[S^f(x)]dV + \int_{V_m}[S^m]dV\right] \tag{4.41}$$

In the iso-stress assumption, it is assumed that the stresses are uniform throughout the unit cell when subject to a homogeneous boundary condition of constant surface tractions as defined in equation (4.9). Similarly, a set of closed form expressions for the effective compliance constants can be obtained.

The 3D fabric geometry model, initially developed by Ko and Chou (1989) to study the compressive behaviour of braided metal-matrix composites, was used by Vandeurzen et al (1996a, 1996b and 1998) to develop 3D elastic models for woven fabric composites. In the fabric geometry model, different yarn systems in a macroscopic unit cell are defined according to the yarn orientation, and each yarn system is treated as a unidirectional lamina. By assuming that all yarn systems have the same strains, i.e., introducing an iso-strain condition in all yarns, the effective stiffness matrix of the composite unit cell can be calculated as the weighted sum of the stiffness matrices of all the yarn systems. Vandeurzen et al (1996a,b) carried out an extensive geometric analysis of woven fabric composites, and then established a macro- and micro-partition procedure to describe even the most complex 2D woven composite structures, with a library of 108 rectangular macro-cells and a library of geometric parameters. The procedure allows definition of the yarn systems in, generally speaking, two ways of micro-partition, as schematically shown in Figure 4.9. In the non-mixed yarn system, the yarn and matrix are modelled separately with the yarns being further partitioned into micro-cells to describe the yarn undulation. In the mixed yarn system, both yarn and matrix are partitioned together to form rectangular micro-cells of mixed yarn system. In the mixed yarn system, fibres of the yarn are redistributed evenly throughout the entire micro-cell with an averaged fibre volume fraction.

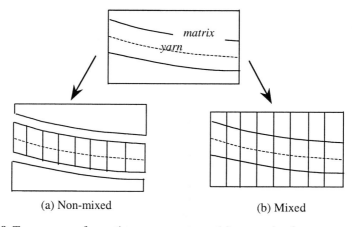

       (a) Non-mixed   (b) Mixed

Figure 4.9 Two ways of creating yarn systems (a) non-mixed yarn systems and (b) mixed yarn systems (Vandeurzen et al, 1996a,b)

Vandeurzen et al (1996b) also considered the iso-stress condition in addition to the iso-strain condition to evaluate the effective properties of the unit cell. For the case of non-mixed yarn systems, the effective properties of the unit cell can be approximated as:

$$[\overline{C}] = \sum_{i=1}^{N_y} w_i [T_i]^T [C_i][T_i] + w_m [C^m]$$

or

$$[\overline{S}] = \sum_{i=1}^{N_y} w_i ([T_i]^T [C_i][T_i])^{-1} + w_m [S^m] \tag{4.42}$$

where w is the volume fraction of the corresponding yarn system or matrix, N_y is the total number of yarn systems and i is the ith yarn system. Subscript m refers to the matrix. The above two equations can be regarded as approximations of equations (4.40) and (4.41) because the transformation matrix $[T]$ is set to be constant in a yarn system.

For the mixed yarn system, above equations can also be used except for $w_m=0$. Vandeurzen et al. (1996b) presented the above equations in a convenient form for implementation in their custom design tool TEXCOMP. They also presented a new model, referred to as the combi-cell model, for mixing up the yarn and matrix as shown in Figure 4.9(b). A combi-cell consists of a yarn layer (Y) and a matrix layer (M) as shown in Figure 4.10(b), which simplifies the micro-cell model in Figure 4.10(a). By minimising complementary strain energy, the effective properties of the combi-cell can be written as

$$[S] = k_Y [A_Y]^T [S_y][A_Y] + k_M [A_M]^T [S_M][A_M] \tag{4.43}$$

where k is the volume fraction and Y and M refer to the yarn and matrix respectively, and $[A_i]$ is the relation matrix, which defines a linear relationship between the externally applied stress $\{\overline{\sigma}\}$ and the layer internal stresses $\{\sigma_i\}$, namely,

$$\{\sigma_i\} = [A_i]\{\overline{\sigma}\} \qquad (i=Y, M) \tag{4.44}$$

The compliance matrix of each micro-cell is then calculated by transforming the compliance matrix of the combi-cell given in equation (4.43) to the unit cell coordinate system.

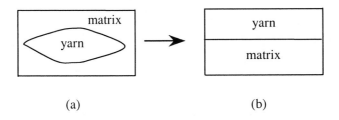

(a)   (b)

Figure 4.10 Combi-cell model (Vandeurzen et al, 1996b)

Tan et al. (1997b) proposed a 3D modelling technique for predicting the linear elastic property of open-packed woven fabrics. Consider the unit cell of an open-packed plain weave as shown in Figure 4.6, introduced is a simplified model as depicted in Figure 4.11. It is clear that the yarn undulation is approximated by a linear inclination. The unit cell is divided into 9 regions which can be represented by the three micro-blocks shown in Figure 4.11(b). Average material properties for each micro block are evaluated first, and then the overall effective properties for the unit cell of the composite are calculated by assembling micro-blocks in the warp direction first into stripes and then in the weft direction or vice versa.

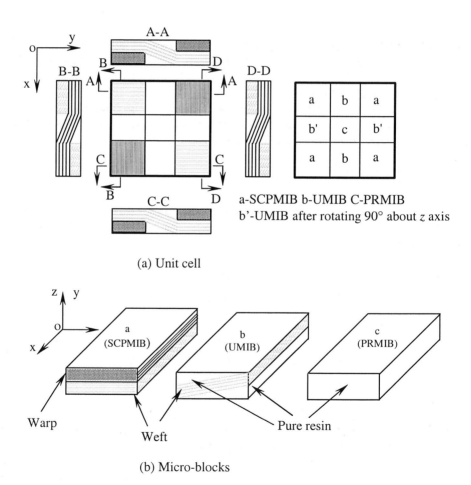

(a) Unit cell

(b) Micro-blocks

Figure 4.11 Simplified models for a unit cell of a plain woven composite (Tan et al, 1997b)

For the micro-block SCPMIB shown in Figure 4.11(b), the individual terms in equation (4.6) are given by:

$$C_{11} = C_{11}^{T}V^{T} + C_{11}^{B}V^{B}, \qquad\qquad C_{12} = C_{12}^{T}V^{T} + C_{12}^{B}V^{B}$$

$$C_{13} = C_{33}\left(\frac{C_{13}{}^{T}V^{T}}{C_{33}{}^{T}} + \frac{C_{13}{}^{B}V^{B}}{C_{33}{}^{B}}\right), \qquad C_{22} = C_{22}^{T}V^{T} + C_{22}^{B}V^{B}$$

$$C_{23} = C_{33}\left(\frac{C_{23}{}^{T}V^{T}}{C_{33}{}^{T}} + \frac{C_{23}{}^{B}V^{B}}{C_{33}{}^{B}}\right) \qquad C_{33} = \frac{C_{33}{}^{T}C_{33}{}^{B}}{V^{T}C_{33}{}^{B} + V^{B}C_{33}{}^{T}} \qquad (4.45)$$

$$C_{44} = \frac{C_{44}{}^{T}C_{44}{}^{B}}{V^{T}C_{44}{}^{B} + V^{B}C_{44}{}^{T}} \qquad\qquad C_{55} = \frac{C_{55}{}^{T}C_{55}{}^{B}}{V^{T}C_{55}{}^{B} + V^{B}C_{55}{}^{T}}$$

$$C_{66} = C_{66}^{T}V^{T} + C_{66}^{B}V^{B}$$

where V^{T} and V^{B} are, respectively, the volume fractions for the top fibre ply (i.e., warp yarn in Figure 4.11(b)) and bottom fibre ply (i.e., weft yarn in Figure 4.11(b)), $C_{ij}{}^{T}$, $C_{ij}{}^{B}$ and C_{ij} are, respectively, the stiffness constants for the top fibre ply, bottom fibre ply and the micro-block SCPMIB.

For the micro block UMIB shown in Figure 4.11(b), the stiffness constants C_{ij} under its local coordinate system can be evaluated using equations (4.45), and then its stiffness constants C_{ij}' under the global coordinate system can be obtained by:

$$[C_{ij}{}'] = [T]^{-1}[C_{ij}][T] \qquad (4.46a)$$

where $[C_{ij}]$ is the stiffness matrix referred to the local coordinate system and $[C_{ij}{}']$ is the stiffness matrix in the global coordinate system. $[T]$ is the Hamiltonian tensor transformation matrix, namely

$$[T] = \begin{bmatrix} l_1^{2} & m_1^{2} & n_1^{2} & 2m_1n_1 & 2l_1n_1 & 2l_1m_1 \\ l_2^{2} & m_2^{2} & n_2^{2} & 2m_2n_2 & 2l_2n_2 & 2l_2m_2 \\ l_3^{2} & m_3^{2} & n_3^{2} & 2m_3n_3 & 2l_3n_3 & 2l_3m_3 \\ l_2l_3 & m_2m_3 & n_2n_3 & m_2n_3+m_3n_2 & l_2n_3+l_3n_2 & l_2m_3+l_3m_2 \\ l_1l_3 & m_1m_3 & n_1n_3 & m_1n_3+m_3n_1 & l_1n_3+l_3n_1 & l_1m_3+l_3m_1 \\ l_1l_2 & m_1m_2 & n_1n_2 & m_1n_2+m_2n_1 & l_1n_2+l_2n_1 & l_1m_2+l_2m_1 \end{bmatrix} \qquad (4.46b)$$

where $l_i = cos(i, x)$, $m_i = cos(i, y)$ and $n_i = cos(i, z)$ for $i=1,2,3$.

When micro-blocks are assembled in the warp or x direction to form the warp stripes, the average properties for a warp stripe can be given by:

$$C_{11}^{S} = \frac{C_{11}^{A}C_{11}^{B}}{N^{A}V^{A}C_{11}^{B} + N^{B}V^{B}C_{11}^{A}} \qquad C_{12}{}^{S} = C_{11}{}^{S}\left[\frac{N^{A}C_{12}{}^{A}V^{A}}{C_{11}{}^{A}} + \frac{N^{B}C_{12}{}^{B}V^{B}}{C_{11}{}^{B}}\right]$$

$$C_{13}{}^{S} = C_{11}{}^{S}\left[\frac{N^{A}C_{13}{}^{A}V^{A}}{C_{11}{}^{A}} + \frac{N^{B}C_{13}{}^{B}V^{B}}{C_{11}{}^{B}}\right] \qquad C_{22}^{S} = N^{A}C_{22}^{A}V^{A} + N^{B}C_{22}^{B}V^{B}$$

$$C_{23}^S = N^A C_{23}^A V^A + N^B C_{23}^B V^B \qquad\qquad C_{33}^S = N^A C_{33}^A V^A + N^B C_{33}^B V^B$$

$$C_{44}^S = N^A C_{44}^A V^A + N^B C_{44}^B V^B \qquad\qquad C_{55}{}^S = \frac{C_{55}{}^A C_{55}{}^B}{N^A V^A C_{55}{}^B + N^B V^B C_{55}{}^A}$$

$$C_{66}{}^S = \frac{C_{66}{}^A C_{66}{}^B}{N^A V^A C_{66}{}^B + N^B V^B C_{66}{}^A} \tag{4.47}$$

where N^A and N^B are the number of micro-blocks A and B within a strip, respectively, V^A and V^B are the volume fractions of a micro-block A and micro-block B, respectively, C_{ij}^A, C_{ij}^B, C_{ij}^S are the stiffness constants for micro-block A, micro-block B and a strip, respectively.

When the micro-blocks are assembled in the weft or y direction to form the weft stripes, equations for the average properties for a weft stripe can be obtained by exchanging 1 and 4 with 2 and 5, respectively, in equation (4.47).

The overall effective properties of the unit cell can be calculated by assembling the warp or weft stripes via employing the equations for properties of the weft or warp stripes, respectively.

The 3D model proposed by Tan et al (1997b) can be extended to take into account the fibre undulation by employing a large number of micro-blocks.

4.3.4 Applications of Finite Element Methods

Finite element methods (FEM) have been used almost universally during the past forty-five years to solve very complex structural engineering problems (Zienkiewicz and Taylor, 1989). When applied to characterise textile composites, FEM visualises them as an assemblage of unit cells interconnected at a discrete number of nodal points. The unit cell is a periodic square array of fibres embedded regularly in the matrix. Hence, if the force-displacement relationship for an individual unit cell is known, it is possible, by using various well-known theories and techniques of elasticity theory, to evaluate its mechanical properties and study the mechanical behaviour of the assembled composite structure.

The general procedure to predict the mechanical properties of a textile composite using FEM consists of 1) dividing the textile composite structure into a number of unit cells and analysing the mechanical properties of a unit cell using FEM, and 2) reconstructing the entire reinforcement geometry by assembling the unit cells for predicting mechanical properties of textile composites. Thus, the ability of a FEA model to predict mechanical properties depends upon the accuracy of modelling the fibre geometry in a unit cell. For the theoretical method, analytical models for elastic properties of composites are generally developed based on classical laminate theory and rule of mixture. Tan et al. (1997a) provided an overview on modelling of mechanical properties of textile composites using the finite element method.

Whitcomb (1989) analysed plain weave composites using 3D finite element analysis, and studied the effect of tow waviness on the effective moduli, Poisson's ratio and internal strain distributions. It was found that the in-plane moduli decreased almost linearly with increasing tow waviness, which was found to create large normal and shear strain concentrations in the composites when subject to a uniaxial loading.

A finite element model for plain weave textile composites was proposed by Glaessgen et al. (1994). The yarns forming a unit cell were considered to be elastic bodies interacting with one another and subject to external loads. The centre line of each yarn was represented as a Bezier curve interpolating a set of discrete support point. The cross-sectional shapes were assumed to be elliptical. The constituent properties of textile composites were approximated as transversely isotropic yarns in an isotropic matrix. The quadratic tetrahedral element in ABAQUS was used. Subsequently, Glaessgen et al. (1996) proposed a method using the textile geometry model combined with the FEM for studying the internal details of displacement, strain, stress and failure parameters. In this method, the geometrical and mechanical modelling was carried out on the structural level of the unit cell.

Since the microstructure of the textile composite is very complex, it is almost impractical to incorporate all architecture parameters in a simple finite element model. In order to obtain reasonable predictions of mechanical properties with a minimum analysis effort, there is a need for more computational efficient methods for performing the analysis (Whitcomb, 1991; Whitcomb and Woo, 1993a).

A 2D to 3D global/local finite element analysis method has been developed by Thompson and Criffin (1990, 1992) to determine interlaminar stress fields for composite laminates with a hole under a remotely loaded uniform uniaxial load. The initial approach used was a 2D global finite element analysis on the whole body or global region with 2D plate finite elements, followed by a more detailed 3D local finite element analysis performed on the local areas of interest with 3D finite elements. The appropriate displacements of the global/local interfaces from 2D global model were applied to the edges of 3D local model. A new finite element analysis approach, called global/local (or macro/micro) analysis method, for textile composite was subsequently proposed for improving analysis efficiency. It was suggested that an accurate global analysis, which determines the local effect on a gross scale, be carried out first, and then as many local analyses as required be conducted.

Whitcomb (1991) proposed an iterative global/local finite element analysis method. The basic idea of this method is that a coarse global model can be employed to obtain displacements or forces that can be used as appropriate boundary conditions for local regions. There may be a potential problem due to the differences in the stiffness of the global and local models. This method was subsequently used for performing linear analysis and geometrically non-linear analysis (Whitcomb and Woo, 1993a,b).

A new type of finite element, referred as macro element, was proposed, in which the tow path was assumed to be sinusoidal (Woo and Whitcomb, 1994). The displacement field within the macro element was assumed to be single field. Because of this assumption, the stresses or strains calculated within the macro element may not be accurate. A new finite element method, referred to as global/local methodology, was proposed. This method is based on 3 types of special macro elements, referred to as coarse microstructure transitional microstructure and fine microstructure. A transitional microstructure is a structure that stands between the coarse and fine microstructures, and a special finite element is needed. It was reported that the predictions obtained using the conventional FEM and the global/local method were in poor agreement when near the global/local boundary.

A 3D finite element model was proposed by Chapman and Whitcomb (1995) to investigate the effect of the assumed tow architecture on the moduli and stresses for plain weaves. In this model, a yarn is assumed to have a sinusoidal tow path and a lenticular cross-section, and macroscopically homogeneous in-plane extension and shear and transverse shear loadings are considered.

A unified prediction method ranging from micro model (named as fibre bundle model) to macro model (named as weaving structure model) was developed by Fujita et al. (1995). In the fibre bundle model, one fibre bundle was modelled with beam elements. Resin elements are set up to connect fibre beam elements. As the rule of mixture is used to calculate the material constants of the fibre element, this model can be used to study micro phenomena within fibre bundle. For the weaving structure model, the weaving structure of textile composites is modelled by connecting the beam elements. Resin existing between crossing fibre bundles is also modelled by resin elements. The section of the fibre bundle is approximately in a rectangular shape, whose area is assumed to be equal to that of the fibre bundle measured. This model was only used to simulate the mechanical behaviours of 2D textile composites, and dependence of the mechanical properties on the textile structural parameters was however not investigated.

4.4 MODELS FOR 3D WOVEN COMPOSITES

In this section, we discuss several modelling schemes for typical 3D woven composites as shown in Figure 4.12. There exist a variety of modelling schemes available. However, our focus will be on the following selected three modelling schemes: orientation averaging models, mixed iso-stress and iso-strain models, and finite element applications.

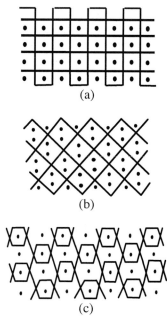

(a)

(b)

(c)

Figure 4.12 Types of 3D woven fabrics, (a) 3D orthogonal interlock, (b) 3D Through-thickness angle interlock, and (c) Layer-to-layer interlock (Tan, 1999)

4.4.1 Orientation Averaging Models

Simple orientation averaging models were originally developed for calculating macroscopically averaged elastic properties of fibre reinforced composites (Tarnopol'skii et al 1973; Kregers and Melbardis, 1978). In these models, the composite is treated as an assemblage of small volumes. In each individual volume, all fibres are aligned and orientated depending on the reinforcement architecture. Each volume can be modelled as unidirectional composites with transversely isotropic properties. The overall effective properties of the composites can then be determined by averaging the response of a representative body to the externally applied loads under the assumption of either uniform stresses or uniform strains. It is clear that the assumption of uniform stresses or strains is identical to that used in the 3D model for 2D woven composites. It may be viewed, to certain extent, as rules of mixtures in the three dimensional case, similar to those in Section 4.2.3.

Cox and Dadkhah (1995) applied the orientation averaging method to 3D woven interlock composite, i.e., layer-to-layer and through-the-thickness angle interlock and orthogonal interlock weaves. For orientation averaging, each composite is divided into stuffer, filler, and two warp weavers volumes with fraction c_i of the total composite volume ($i=s, f, w_1$ and w_2 for stuffer, filler, and two warp weavers volume) with a sum of c_i being unity. Similar to equation (4.42), the following approximate expression for the stiffness matrix of 3D woven composites with ideal geometry is obtained:

$$[\overline{C}] = \sum_{i=s,f,w_1,w_2} w_i [T_i]^T [C_i][T_i] \qquad (4.48)$$

It was found that the ideal geometry is far different from the true geometry. Both stuffer and filler yarns are not straight and there exists significant out-of-plane waviness, which varies along the stuffer and filler directions. To take into account the most important effect of tow waviness on elastic properties, a symmetrical normal distribution is formed for the out-of-plane alignment angle, ξ, as follows:

$$F_\xi(\xi) = \int_{-\infty}^{\xi} f_\xi(\xi') d\xi'$$

with the density function given by

$$f_\xi(\xi) = \frac{1}{\sigma_\xi \sqrt{2\pi}} e^{-\xi^2/2\sigma_\xi^2}$$

where σ_ξ represents the width of the distributions. A waviness knockdown factor is defined as

$$\eta \approx 1 - \sigma_\xi^2 \left[\frac{E_1}{G_{12}} - 2(1+\nu_{12}) \right] \qquad \text{for } \sigma_\xi \leq 10° \qquad (4.49)$$

The waviness knockdown factor is used to reduce the values of Young's modulus in the tow direction and Poisson ratio ν_{12} for the stuffer and filler yarns.

As found by Cox and Dadkhah (1995), the orientation averaging model with simple corrections for tow waviness can provide an excellent prediction of the in-plane macroscopic elastic constants and a fair estimation for elastic constants related to through-thickness strains.

4.4.2 Mixed Iso-Stress and Iso-Strain Models

Tan et al. (1998, 1999a,b) proposed a mixed iso-stress and iso-strain based unit cell modelling scheme for predicting mechanical and thermo-elastic properties for 3D orthogonal and angle-interlock composite materials. The modelling scheme was experimentally validated by comparing the measured elastic properties of 3D orthogonal carbon fibre reinforced composites and 3D glass fibre reinforced composites with those predicted (Tan et al., 2000b, 2001). In the following, we will describe the fundamentals of the mixed iso-stress and iso-strain unit cell modelling scheme by considering a 3D orthogonal woven composite material.

   Consider a piece of material from a 3D orthogonal woven composite as shown in Figure 4.13, in which three types of yarns (i.e. warp, weft and z yarns) of assumed rectangular cross-sectional shapes are placed in three mutually orthogonal directions. The marked volume can be treated as a unit cell as shown in Figure 4.14.

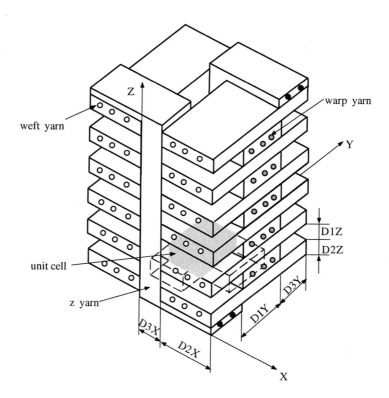

Figure 4.13 A schematic of idealised 3D orthogonal woven preform (resin removed from the preform) (Tan et al, 1998, 1999a,b)

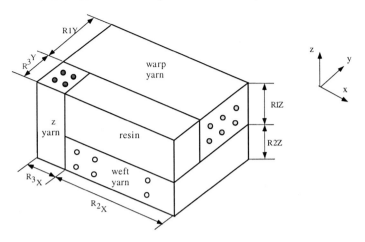

Figure 4.14 Schematic of idealised 3D orthogonal woven fabric unit cell (Tan, 1999)

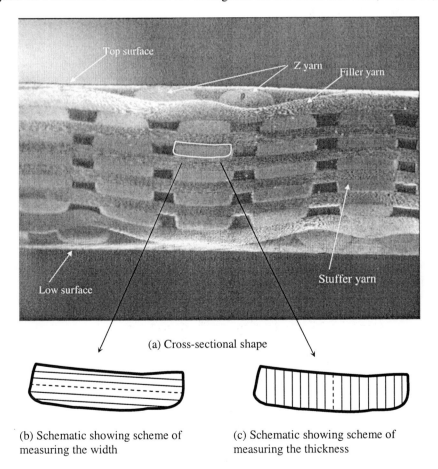

(a) Cross-sectional shape

(b) Schematic showing scheme of
measuring the width

(c) Schematic showing scheme of
measuring the thickness

Figure 4.15 micrograph of the cross-sectional shape of a 3D orthogonal woven CFRP composite (Tan et al, 2000a,b)

As shown in Figure 4.15, there exists a remarkable difference between the idealised and true geometry in a 3D orthogonal woven composite material. The procedure to determine all idealised geometrical dimensions was detailed in Tan et al (2000b). Figure 4.15 schematically shows the scheme used to determine the cross-sectional dimensions of a rectangular shaped stuffer yarn.

The unit cell shown in Figure 4.14 can be subdivided into four blocks of cubic shape by cutting through two planes that are perpendicular to each other and pass through the selected two interfacial planes between the three yarns and the resin. For example, one cutting plane is selected as the interfacial surface between the warp and weft yarns parallel to the *xy* plane, and the other one as the interfacial plane between the warp yarn and the resin parallel to the *xz* plane. In this case, the four blocks are shown in Figure 4.16. It is clear that the overall properties of the unit cell can be evaluated by estimating the properties of each block. Figure 4.17 depicts all three possible ways of assemblage of each block comprising of two constituent sub-blocks A and B.

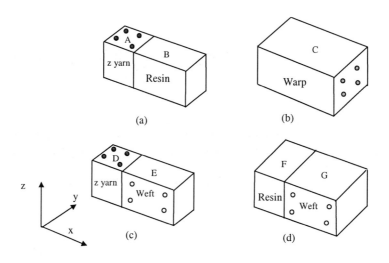

Figure 4.16 an illustrative example of the 4 blocks of 3D orthogonal woven composite material (Tan et al, 1999a,b)

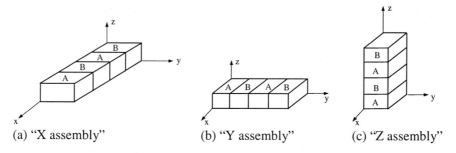

Figure 4.17 Three possible assemblages of block with A and B constituent sub-blocks (Tan et al, 1999a,b)

As shown in Figure 4.17, an "X assembly" is assembled with N^A blocks of material A and N^B blocks of material B in the x direction. The overall material properties are given via the following equations:

$$C_{11}{}^S = \frac{C_{11}{}^A C_{11}{}^B}{N^A V^A C_{11}{}^B + N^B V^B C_{11}{}^A},$$

$$C_{12}{}^S = C_{11}{}^S \left[\frac{N^A C_{12}{}^A V^A}{C_{11}{}^A} + \frac{N^B C_{12}{}^B V^B}{C_{11}{}^B} \right],$$

$$C_{13}{}^S = C_{11}{}^S \left[\frac{N^A C_{13}{}^A V^A}{C_{11}{}^A} + \frac{N^B C_{13}{}^B V^B}{C_{11}{}^B} \right],$$

$$C_{22}{}^S = N^A C_{22}{}^A V^A + N^B C_{22}{}^B V^B,$$

$$C_{23}{}^S = N^A C_{23}{}^A V^A + N^B C_{23}{}^B V^B,$$

$$C_{33}{}^S = N^A C_{33}{}^A V^A + N^B C_{33}{}^B V^B,$$

$$C_{44}{}^S = N^A C_{44}{}^A V^A + N^B C_{44}{}^B V^B,$$

$$C_{55}{}^S = \frac{C_{55}{}^A C_{55}{}^B}{N^A V^A C_{55}{}^B + N^B V^B C_{55}{}^A},$$

$$C_{66}{}^S = \frac{C_{66}{}^A C_{66}{}^B}{N^A V^A C_{66}{}^B + N^B V^B C_{66}{}^A} \tag{4.50}$$

where N^A and N^B are the numbers of micro-blocks A and B within a strip respectively, and V^A and V^B are the volume fractions of a micro-block A and a micro-block B in a strip respectively, $C_{ij}{}^A$, $C_{ij}{}^B$, $C_{ij}{}^S$ are the stiffness constants for a micro-block A, a micro-block B and a strip respectively.

The overall material properties for the Y assembly block are given by the following equations:

$$C_{11}{}^S = N^A C_{11}{}^A V^A + N^B C_{11}{}^B V^B,$$

$$C_{12}{}^S = C_{22}{}^S \left[\frac{N^A C_{12}{}^A V^A}{C_{22}{}^A} + \frac{N^B C_{12}{}^B V^B}{C_{22}{}^B} \right],$$

$$C_{13}{}^S = N^A C_{13}{}^A V^A + N^B C_{13}{}^B V^B,$$

$$C_{22}{}^S = \frac{C_{22}{}^A C_{22}{}^B}{N^A V^A C_{22}{}^B + N^B V^B C_{22}{}^A},$$

$$C_{23}{}^S = C_{22}{}^S \left[\frac{N^A C_{23}{}^A V^A}{C_{22}{}^A} + \frac{N^B C_{23}{}^B V^B}{C_{22}{}^B} \right],$$

$$C_{33}{}^S = N^A C_{33}{}^A V^A + N^B C_{33}{}^B V^B,$$

$$C_{44}{}^S = \frac{C_{44}{}^A C_{44}{}^B}{N^A V^A C_{44}{}^B + N^B V^B C_{44}{}^A},$$

$$C_{55}{}^S = N^A C_{55}{}^A V^A + N^B C_{55}{}^B V^B,$$

$$C_{66}{}^S = \frac{C_{66}{}^A C_{66}{}^B}{N^A V^A C_{66}{}^B + N^B V^B C_{66}{}^A}, \tag{4.51}$$

The equations required to calculate the overall material properties for the Z assembly are given as follows:

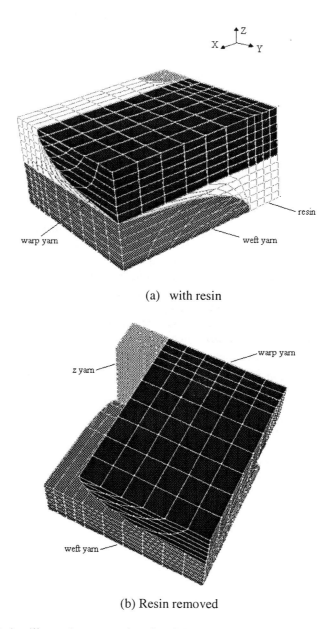

(a) with resin

(b) Resin removed

Figure 4.18 An illustrative example of a full 3D FE model of a unit cell to a 3D orthogonal woven composite material (Tan, 1999)

To evaluate the overall properties for a 3D woven composite material, six independent displacement fields should be applied respectively to the finite element model of the unit cell using the Lagrange multiplier method (see Tan et al., 2000b). The six displacement fields yield six independent overall uniform strain fields, e.g., $\varepsilon_x = 0.001$,

$\varepsilon_y = 0.001$, $\varepsilon_z = 0.001$, $\gamma_{xy} = 0.001$, $\gamma_{yz} = 0.001$ and $\gamma_{zx} = 0.001$. The corresponding engineering constants can be determined using equation (4.10).

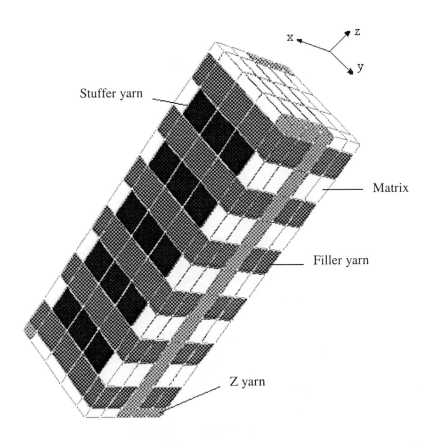

Figure 4.19 FEA model for a simplified 3D orthogonal woven fabric laminate block (Tan, 1999)

4.4.3.2 Binary Models

Cox et al. (1994) and Xu et al. (1995) developed a finite element model known as the "binary model" for simulating 3D woven textile composites, as shown in Figure 4.12, in the elastic regime. As an illustrative example, Figure 4.20 depicts a typical arrangement of nodes on tow and effective medium elements in a small volume of a layer-to-layer angle interlock woven composite. In the binary model, the two-node tow elements represent the axial properties of tows, while the effective medium element represents all other properties of the tow, resin pockets etc in an average sense. This "effective medium" element was considered to be homogeneous and isotropic. The characterisation of the tow, effective medium as well as the spring element plays an important role in predicting both the mechanical properties and failure strength of the angle interlock woven composites. Details of characterisation can be found in Cox et al. (1994) and Xu et al. (1995).

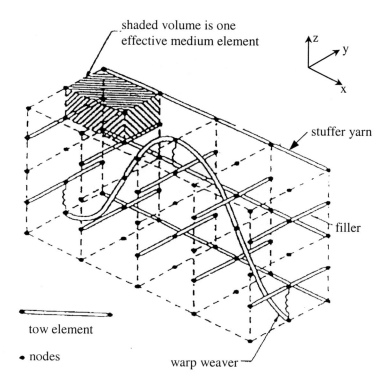

Figure 4.20 Tow and matrix elements in a layer-to-layer angle interlock woven composite (Cox et al, 1994; Xu et al, 1995)

4.5 UNIT CELL MODELS FOR BRAIDED AND KNITTED COMPOSITES

There are several unit cell models that have been developed during the past two decades. This section will present a brief overview of the models available based on the previous review by Tan et al (1997a).

4.5.1 Braided Composites

A self-consistent fabric geometry model was proposed by Pastore and Gowayed (1994) for predicting the elastic properties of textile composites. This model is a modification of the Fabric Geometry Model (Ko and Pastore, 1985), which relates the fibre architecture and material properties of textile reinforced composites to its global stiffness matrix through micromechanics and the stiffness averaging technique. It was reported that the self-consistent fabric geometry model gave good predictions of elastic properties by relating these properties to the constituent material properties and the fibre architecture geometry.

FEM has also been widely applied to investigate the mechanical properties of braided fabrics. In 1986, Ma et al. (1986) proposed a 'diagonal brick model' for 3D braided textile composites as shown in Figure 4.21, which was based on the concept of a simplified fibre unit cell structure. As shown in the figure, it was noted that this model consists of a brick-shaped element of bulk resin with four parallel bar elements along four edges of the brick plus four diagonal bar elements, and the unit cell is centred around an "interlock" of yarns. The spatial orientation of each diagonal bar element in the unit cell is defined by its orientation angle with respect to x-axis as:

$$\theta = \tan^{-1} \frac{\sqrt{P_b^2 + P_c^2}}{P_a} \tag{4.53}$$

The "inclination" and "off-axis" angles α and β are defined as (Whitney and Chou, 1989):

$$\tan\alpha = \frac{\tan\theta}{\sqrt{R^2(1+\tan^2\theta)+1}} \tag{4.54a}$$

$$\tan\beta = \frac{R\tan\theta}{\sqrt{R^2+1}}, \text{ where R=b/c} \tag{4.54b}$$

However, crimping of fibre, which was assumed to occur at the corners of the cell, was neglected. The intersection of fibres at the unit cell centre was also ignored (Whitney and Chou, 1989).

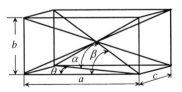

Figure 4.21 Geometrical schematic of a unit cell for a 3D braided composite (Ma et al, 1986)

Yang et al. (1986) introduced a new model called 'fibre inclination model' by extending the lamination theory for predicting the elastic properties of 3D braided composites. This model treats the unit cell of a composite as an assemblage of inclined unidirectional laminae. The laminate approximation of the unit cell structure is shown schematically in Figure 4.22, in which the inclined laminae are composed of yarns in four diagonal directions in the unit cell. The inclination angles of the laminae θ_α and the off-axis angle of yarn segment with respect to the x-axis θ_β are expressed as:

$$\theta_\alpha = \pm \tan^{-1} \frac{P_c}{L} \tag{4.55a}$$

$$\theta_\beta = \pm \tan^{-1} \frac{P_b}{P_a} \tag{4.55b}$$

This approach is an extension of the one-dimension "fibre undulation model" developed by Ishikawa and Chou (1986). No experimental tests were attempted to verify this model, although it was stated that the relevant predictions showed reasonably good agreement with the experimental data obtained by several other researchers. As the fibre volume fraction of 3D braid is normally over 0.5, it maybe possible to model the yarns as bar elements (rather than dimensionless) in the four diagonal directions.

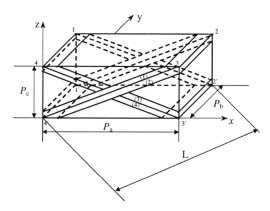

Figure 4.22 Geometrical schematic of a unit cell of the fibre inclination model composed of four unidirectional laminae for braided composites (Yang et al, 1986)

All the approaches described above in modelling braided composites are to define a unit cell geometry for a braided structure without providing any relationship between processing variables and geometric parameters. Hence, these models may not be used to study the optimisation of the braided fabric architecture for their structural applications.

Byun et al (1991) developed a fabric geometric model using lamination analysis and the stiffness averaging method. This model combines the micro-cell model and macro-cell model, and can be utilised to predict the elastic constant of 2-step braided composites. The micro-cell model is constructed for thin specimens so that the two-dimensional approximation of the classical lamination theory can be applied. Given the geometric parameters of the micro-cell, fibre and matrix properties and composite fibre volume fraction, effective in-plane elastic properties can be calculated based on a pointwise application of lamination theory. The compliance (or stiffness) constants are averaged to obtain an average compliance (or stiffness) for the micro-cell. The macro-cell model is developed for the entire cross-section of the specimen. In this model, variations in the braider yarn orientation along its length are taken into account by introducing an average yarn orientation angle. It was reported that good agreement between the model predictions and test results were achieved for axial tensile modulus. A lack of agreement between the predicted and measured shear moduli was reported.

The effects of the braid angle, yarn size and axial yarn content on the mechanical properties were investigated by Naik et al (1994) for 2D triaxially braided textile composites. The numerical results showed that the mechanical properties are more sensitive to variability in braid angle than to variations in axial yarn content, and are not sensitive to change in yarn size. Increasing the braid angle decreases the longitudinal modulus but increases the transverse modulus, and the in-plane shear modulus values peak at the braid angle of $\pm 45°$. An increase in the axial yarn content results in a higher longitudinal modulus, but a lower in-plane shear modulus and Poisson's ratio. The out-of-plane properties remain virtually unchanged with variations in the braid angle and the axial yarn content.

4.5.2 Knitted Composites

As knitted fabrics are not often used in structural applications, and their geometric architectures are more complex compared to woven and braided fabrics, only a limited attention has been given to the modelling of the mechanical properties of knitted fabric composites (Ruan and Chow, 1996).

A simple stiffness model was proposed by Rudd et al. (1990) for predicting the mechanical properties of weft knit glass fibre/polyester laminates. This model was developed based on the well-known rule of mixtures approach. The comparison between the predicted and experimental results suggested that the model require modification to take into account the fabric relaxation.

Ramakrishna and Hull (1994) created an analytic model for predicting the elastic moduli and tensile strengths of knitted fabric laminates. In this model, the reinforcement efficiencies of yarns are incorporated into the rule-of-mixture, and the effects of the out-of-plane yarns are however neglected. The tensile strengths of composites were estimated by the strength of straight resin-impregnated yarns. It was reported that the predicted elastic moduli were in a reasonable agreement with experimental results while a significant discrepancy existed between the experimental and predicted tensile strengths.

Ruan and Chou (1996) developed geometric models for plain-stitch and rib-stitch fabric composites. These models were developed using the yarn configuration and microstructures of the preform observed using an optical microscope. In this analytical model, it is assumed that an infinitesimal segment, which is formed by two parallel planes perpendicular to the warp (loading) direction, is subject to a uniform strain. Modelling of elastic behaviour was conducted using an averaging method. It was reported that the tensile and shear moduli determined by the analytical model were higher than the actual values.

Ko et al. (1986) proposed a fabric geometry model for predicting the tensile properties of the warp-knit fabric composites. This model was developed based on the unit cell concept and laminate theory. It was reported that there was a good agreement between the predicted and experimental test results.

A "cross-over model" was proposed by Ramakrishna (1997a) for expressing the crossing over of looped yarns of knitted fabric. This model considers the three-dimensional orientation of yarns in the knitted fabric composite. Each impregnated yarn is idealised as a curved unidirectional lamina. The effective elastic properties of the yarns were estimated using the laminated plate theory. The elastic properties of the composite were determined by combining the elastic properties of yarns and the resin-rich regions. The analytical model was extended to predict the elastic properties of knitted-fabric composite with different fibre volume fraction (Ramakrishna, 1997b). The predicted tensile properties compared favourably with the experimental results.

4.6 FAILURE STRENGTH PREDICTION

The failure mechanisms and procedures of a 3D textile composite material at the micromechanical scale vary with type of loading and are intimately related to the properties of the constituents, i.e., fibre, matrix and interface-interphase, and the micro architectures of fibre yarns as well as yarn-matrix interphase. Strength predictions are based on micromechanical analyses and point-based failure criteria, and may be accurate with regard to failure initiation at critical points. However, such predictions are only approximate in the context of global failure of the composite material.

Due to the complexity and irregularity of fibre distributions and the limitation of measurement facilities, there is a lack of detailed knowledge and understanding of failure mechanisms for 3D textile composites. It is difficult, if not impossible, to perform thorough micromechanical analyses to obtain reliable strength predictions for 3D textile composites under a general type of loading. For this reason, it may be preferable to carry out the strength predictions by treating fibre yarns as an anisotropic property, comprising of unidirectional fibres and matrix, and embedded in an isotropic matrix. Similar to mechanical property predictions, a textile composite material may be idealised as fibre yarns of different architectures embedded in a matrix. In this approximation, failure may occur in the yarns, matrix and the interfaces amongst yarns and matrix in a textile composite subject to a general type of loading.

The strength of a yarn along an arbitrary direction may be correlated with the basic strength parameters. Similar to a unidirectional lamina, a yarn may be characterized by a number of basic strength parameters with respect to its principal material directions from the macro-mechanical point of view. For example, the first principal material direction, axis 1, is chosen as along the fibre direction or the tangential direction at any point along the yarn centreline path, while the second and third principal material directions, axis 2 and 3, are selected to be two orthogonal axes within a plane perpendicular to the first principal material direction. In general, there are three tensile and three compressive strength parameters in the three principal material directions, and three shear strengths in the mutually orthogonal planes passing through any two principal material axes. However, for a unidirectional yarn, it is desirable to reduce the number of independent strength parameters from nine to six. This is because the tensile and compressive strength parameters in direction 2 can be assumed to be the same as those in direction 3, and the shear strength parameter in the plane going through axis 1 and 2 can be assumed to be identical to that going though axis 1 and 3. When all the basic strength parameters are known, the maximum stress criterion and maximum strain criterion (see Jones, 1975; Christensen, 1979) may be used in conjunction with stresses and strains computed at a critical point in a yarn to predict the failure load. The Tsai-Wu (1971) tensor polynomial criterion may also be used to predict the yarn strength. Failure that occurs in the matrix can be predicted using conventional failure criteria for a homogeneous and isotropic material. Failure that occurs along the interface between two yarns may be predicted using those failure criteria for predicting interlaminar delamination in composite laminates.

Tan et al (2000a,b) investigated the failure of the 3D orthogonal woven composite specimen in tensile loading. They carried out a full 3D finite element analysis of a unit cell for the 3D orthogonal woven composite by modelling yarns as a homogeneous and orthotropic property and matrix as being isotropic. Maximum stress criterion and the rule of mixture were employed to predict the tensile strengths in both in-plane directions, i.e., the stuffer yarn and weft yarn directions. A good correlation was noted for the tensile strength in the direction of the stuffer yarns. Due to waviness of the weft or filler yarns (see Figure 4.23a), there is a remarkable difference between the measured and predicted strength. Tan et al. (2001) also investigated the mechanical properties and failure mechanisms of 3D orthogonal woven E-glass/epoxy composite materials. Their results show that there is a reasonably good correlation between the measured tensile strengths and those predicted using the rule of mixture. Callus et al (1999) performed tensile tests of glass fibre reinforced polymer composites with 3D orthogonal, normal layered interlock, and offset layered interlock woven architectures.

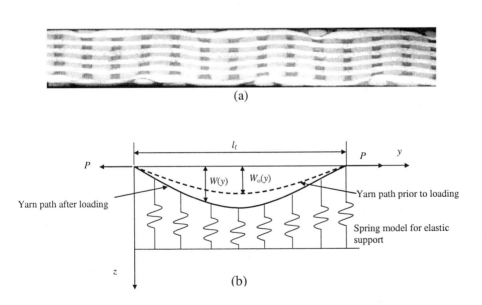

Figure 4.23 (a) micrograph of cross-section showing waviness of filler yarns, and (b) Schematic of the curved beam model (Tan et al, 2000a; Tong et al, 2002)

To take into account the yarn waviness in strength prediction, Tong et al (2002) modelled each individual weft yarns between any two adjacent stuffer yarns as a beam on elastic foundation with an initial deflection as shown in Figure 4.23b. The initial deflection is introduced to model the filler yarn waviness, which can be experimentally measured from a micrograph of the cross-section. The cross-sectional micrographs (see Figure 4.15) show that the yarn waviness can be approximated by a sine function. Possible failure mode range and loci were then predicted using the model. A good agreement between the measured and predicted failure load ranges and the observed failure loci was reported. This will be discussed in details in Section 5.3.1.

Emehel and Shivakumar (1997) proposed a tow collapse model for predicting compressive strength of multiaxial laminates and textile composites. In this model, the unidirectional composite compressive strength model based on microbuckling of fibres embedded in a rigid-plastic matrix was employed to develop a resulting expression in terms of the matrix yield strength under the fibre constraint, fibre tow inclination angle, fibre volume fraction, and the area fractions of various sets of inclined tows. The predicted strengths agree reasonably well with those measured in the experimental tests.

Chapter 5

3D Woven Composites

5.1 INTRODUCTION

3D woven composite is a new type of advanced engineering material that is currently used in only a few niche applications. The most significant applications are stiffeners for the air induct duct panels on the Joint Strike Fighter, aircraft wing joints on the Beech Starship, and rocket nose cones. 3D woven sandwich composite reinforced with distance fabric is also used in modest amounts, such as floor panels for trains, hard-tops for convertible cars, and the deck and top-side structure for a fishing boat. While the present use of 3D woven composite is limited, the potential use is impressive and wide-ranging with various possible applications in the aerospace, marine, infrastructure, military and medical fields. As described in chapter 1, in the future this composite may be used in a diverse variety of items ranging from jet-engine components to personnel body armour to artificial limbs.

While the future of 3D woven composites appears promising, it is not assured. Many challenges are facing the increased application of this material. A major factor is that the cost of 3D woven composites is currently higher than 2D prepreg or fabric laminates for many applications. It was discussed in chapter 2 that the 3D weaving process has the potential to reduce lay-up and assembly costs in fabrication, however 3D woven fabric is not yet produced in large commercial quantities at low cost. Another impediment to the increased use of 3D woven composite in the aircraft industry is the high cost of certifying these and other new materials for primary load-bearing structures. Until the cost savings and other benefits of 3D woven composite are fully appreciated, then aircraft manufacturers will continue using conventional 2D laminates in most composite components.

Another significant challenge is that many designers, fabricators and users of composites are unsure of the potential benefits of using 3D woven composite. Most sectors of the composite industry do not fully appreciate the benefits gained from using 3D woven material, such as reduced fabrication cost, greater design flexibility, improved impact resistance, and superior through-thickness mechanical properties. The design and fabrication of 3D woven composite is described in Chapter 2 and micromechanical models for predicting their stiffness and strength are outlined in Chapter 4. In this chapter the in-plane mechanical properties, delamination resistance and impact damage properties of 3D woven composites are described. In Section 5.2 the microstructural features of 3D woven composites that affect the mechanical and impact properties are described. This includes a description of microstructural damage such as fibre crimping, fibre damage and z-binder distortion that degrade the in-plane and through-thickness properties. The mechanical properties and failure mechanisms of 3D woven composites under tension, compression, bending, interlaminar shear and fatigue loads are described in Section 5.3. Following this, the delamination resistance and interlaminar fracture toughening mechanisms are outlined in Section 5.4 and the impact damage tolerance in Section 5.5. The properties of 3D woven sandwich composites made using distance fabric are given in Section 5.6.

5.2 MICROSTRUCTURAL PROPERTIES OF 3D WOVEN COMPOSITES

The microstructure of a 3D woven composite is determined largely by the fibre architecture to the woven preform and weaving process, and to a lesser extent by the consolidation process.   Various types of microstructural defects are inadvertently produced during 3D weaving that can degrade the in-plane, through-thickness and impact properties. The main types of defects are abrasion, breakage and distortion of the in-plane and z-binder yarns as well as resin-rich and resin-starved regions.

Abrasion and breakage of the warp, weft and z-binder fibres[1] are common types of damage incurred in 3D weaving that are difficult to avoid. This damage occurs by the bending of yarns in the weaving process and as yarns slide against the loom machinery (Lee et al., 2001, 2002). For example, Figure 5.1 shows broken filaments in a yarn that is passing through the guide to a 3D weaving loom. Figure 5.2 shows fragments of broken fibre caused by 3D weaving. This damage from the weaving process can cause a large reduction to the tensile strength of brittle yarns. Figure 5.3 shows cumulative probability distribution plots by Lee et al. (2002) of the failure strength of an E-glass yarn after different stages of weaving. It is seen that the tensile strength decreases progressively after the tensioning, warping and take-up stages, causing an overall strength reduction of about 30%. The loss in yarn strength is dependent on a number of factors, such as the yarn diameter, 3D fibre architecture, and type of loom. It is also strongly influenced by the brittleness of the fibre, with glass yarns experiencing a greater loss in strength than carbon or Kevlar yarns. It is worth noting that the fibre damage and loss in strength shown here for 3D woven fabric is also experienced with 2D fabric during conventional (single-ply) weaving.

In addition to abrasion and fracture, the fibres are distorted and crimped by 3D weaving. The warp and weft yarns in 3D woven preforms have a large amount of waviness, and typically the fibres are misaligned from the in-plane direction by 4 to 12° (Cox et al., 1994, Callus et al., 1999; Kuo and Ko, 2000). In extreme cases, the misalignment can be greater than 12°, particularly in fibre segments close to the z-binders.   The fibres in 3D preforms show much greater waviness than in 2D prepreg laminates, where the waviness is under 2-3°. The fibres in 3D preforms also experience extreme localised distortion, known as crimping, at the surface regions where the z-binder yarns cross-over the in-plane tows. The crimping of a filler tow is shown schematically in Figure 5.4. This pinching by the z-binder crimps the surface yarns, thus causing them to collimate (or bunch together) which creates pockets rich in resin between them.

The z-binder yarns can also experience excessive distortion in 3D woven composites. This distortion can occur by a high tensile force applied to the z-binder in the weaving process, as discussed earlier in Chapter 2. It can also occur during

---

[1] Different terminology is used to describe the fibres in 3D woven composites. The warp yarns are also known as 'load-bearing yarns' or 'stuffers' while weft yarns can be called 'transverse yarns' or 'fillers'. The z-binder yarn is also known as a 'weaver'.

consolidation when excessive overpressure can squash the preform and thereby misalign the z-binders. Figure 5.5 illustrates the distortion to a z-binder yarn in a 3D orthogonal composite, resulting in a quasi-sine-wave path. Studies by Callus et al. (1999) and Leong et al. (2000) report misalignment angles for z-binder yarns of up to ~45° from the square-wave profile expected in 3D orthogonal composites.

Figure 5.1 Broken fibres caused by 3D weaving.

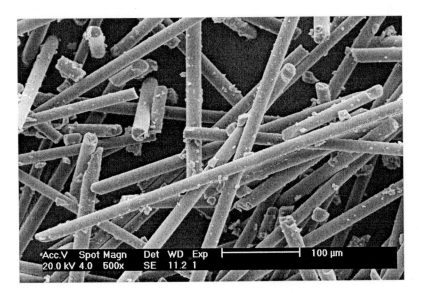

Figure 5.2 Fragments of broken glass fibres caused by 3D weaving.

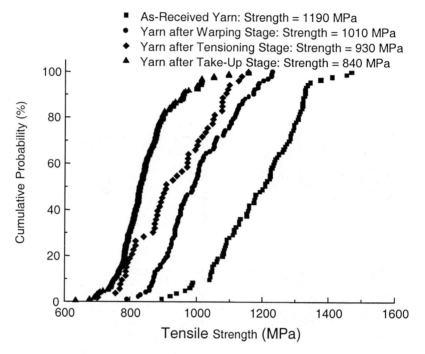

Figure 5.3 Cumulative probability distribution plots of the tensile strength of a 300 tex E-glass yarn determined after different stages of the 3D weaving process. The average tensile strength value of the yarn after each weaving stage is shown (Adapted from Lee et al., 2002).

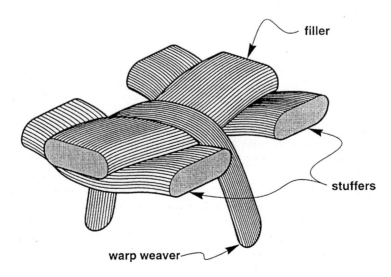

Figure 5.4 Schematic of crimping of a surface tow (or filler) by a z-binder yarn (or warp weaver) Cox et al. (1994).

The squashing of warp, weft and z-binder yarns creates regions of high-fibre content in 3D preforms. When the preforms are consolidated, viscous resins can have difficulty infiltrating these regions that can lead to porosity. 3D woven preforms also have localised regions of low fibre content, particularly where the in-plane yarns have been crimped and pushed aside by the z-binders. Upon consolidation these regions become rich in resin (Farley et al., 1992; Leong et al., 2000).

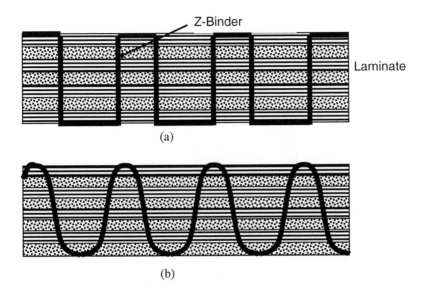

(a)

(b)

Figure 5.5 (a) Idealised and (b) actual profiles of a z-binder yarn in a 3D orthogonal composite. The z-binder is supposed to have a square-wave profile, but in reality can be distorted into a quasi-sinusoidal profile.

As another illustrative example, Figure 5.6 also depicts a 3D orthogonal woven composite that comprises of stuffer yarns, filler yarns and z-binders of nominal proportions of 1:1.2:0.2 (Tan et al, 2000a). The overall fiber volume fraction for the 3D orthogonal woven composite panels is 43%. The 3D orthogonal woven composite panels have an average thickness of 2.57 mm. Figure 5.6(b) depicts a micrograph of the cross section A-A as shown in Figure 5.6(a). There are six filler yarn layers and five stuffer yarn layers. It is clear that all filler yarns are not straight. The misalignment of the internal filler yarns appear to be less severe than that of the two surface filler yarns. Figure 5.6(b) also shows that the cross section of the stuffer yarns is only slightly distorted from its ideal rectangular shape. Figure 5.6(c) shows the micrograph of the cross section B-B as indicated in Figure 5.6(a). It is clearly demonstrated that the z-binder exhibits a smooth periodically curved shape rather than an idealised rectangular shape. The cross sectional shape of all four inner filler yarns appears to be close to a skewed rectangle, and that of the two surface filler yarns is severely distorted from a rectangle into a skewed triangle or quadrilateral.

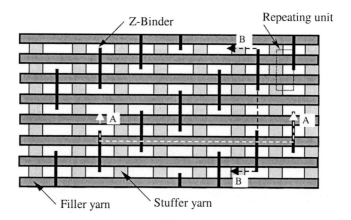

(a) A schematic of the top view for the 3D orthogonal woven CFRP composite

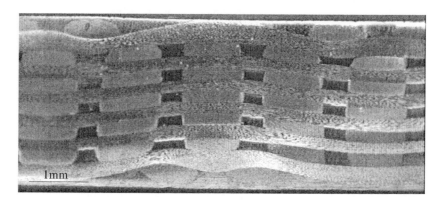

(b) Micrograph of cross section A-A showing misalignment of filler yarns

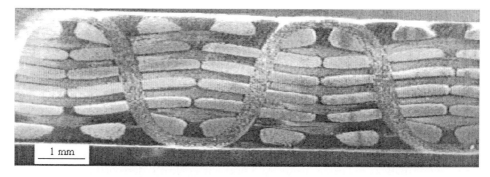

(c) Micrograph of cross section B-B showing true path of the z-binder and distorted filler yarns

Figure 5.6 Architectural features of a 3D orthogonal woven CFRP composite (Tan et al, 2000a,b)

5.3 IN-PLANE MECHANICAL PROPERTIES OF 3D WOVEN COMPOSITES

5.3.1 Tensile Properties

The tensile properties and failure mechanisms of 3D woven composites have been investigated since the mid-1980s, but only recently has an understanding of their tensile performance began to emerge. Tensile studies have been performed on 3D woven composites with orthogonal or interlock fibre structures made of carbon, glass or Kevlar. Numerous studies have compared the tensile properties of 3D woven composites against 2D laminates with a similar (but not always the same) fibre content, and different results are reported. The Young's modulus of some 3D woven composites is lower than the modulus of their equivalent 2D laminate. This difference is shown by a comparison of tensile stress-strain curves for a 2D and 3D woven composite in Figure 5.7. This data from Lee et al. (2002) shows that the Young's modulus of the 3D composite is about 35% lower than the 2D laminate. Other tensile studies also report that the Young's modulus of a 3D woven composite is lower than a 2D laminate, with the reduction ranging from ~10% to 35% (Ding et al., 1993; Guess and Reedy, 1985). However, in some cases the tensile modulus of the 3D woven composite can be slightly higher than the 2D laminate (Arendts et al., 1989; Chen et al., 1993).

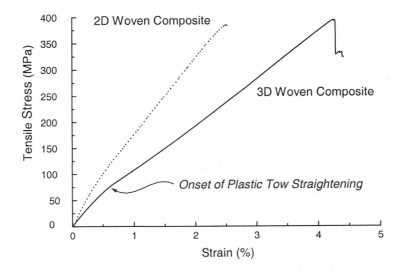

Figure 5.7 Tensile stress-strain curves for a 2D and 3D woven composite.

The Young's modulus values for a variety of 3D woven composites are plotted against their z-binder content in Figure 5.8. In this figure the Young's modulus of the 3D woven composite is normalised to the modulus of the equivalent 2D laminate. It is important to note, however, that the 2D laminate is not always exactly equivalent because the fibre contents of the 3D and 2D composites being compared are rarely the same, and often differ by several percent. With the exception of a few outlying values, it is seen in Figure 5.8 that the Young's modulus of a 3D composite is always within 20% of the modulus of the 2D laminate. Only rarely is the stiffness of a 3D composite higher or lower by more than 20%. Figure 5.8 also shows that the Young's modulus of a 3D woven composite is not influenced significantly by the z-binder content or fibre structures (ie. orthogonal vs. interlock). The reason for the higher Young's modulus of some 3D woven composites is probably due to a slightly higher fibre content than the 'equivalent' 2D laminate. The lower modulus of the other 3D woven composites is due to higher fibre waviness of the load-bearing yarns caused by the z-binder.

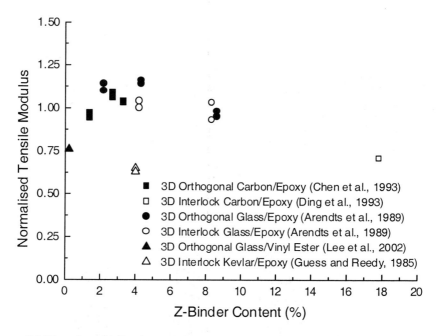

Figure 5.8 Plot of normalised Young's modulus against z-binder content for various 3D woven composites.

The micromechanical models described in Chapter 4 can be used to accurately determine the Young's modulus of 3D woven composites. Even the simplest models, such as the rule-of-mixtures, provide good estimates of modulus. As an example, Figure 5.9 gives a comparison of the measured modulus for different types of 3D woven composite against the theoretical modulus calculated using rule-of-mixtures. It is seen that the agreement between the experimental and theoretical modulus values are within 10% in all cases.

Tan et al (2000a) measured the in-plane Young's moduli and Poisson's ratio for the 3D orthogonal woven CFRP composites as shown in Figure 5.6. Table 5.1 gives a comparison between the measured in-plane elastic constants and those predicted using the block laminate and the unit cell models presented in Chapter 4. In Table 5.1, E_1 and E_2 are the Young's modulus in the stuffer and filler direction respectively, while v_{12} is the Poisson's ratio. The experimental results and those predicted using the laminate block models are detailed in Tan et al (2000a,b). The modulus in the filler yarn direction is larger than that in the stuffer yarn direction because the fibre content in the filler yarn direction is 20% more than that in the stuffer yarn direction. The predicted results for the unit cell model are taken from Kim et al (2001), who model the full 3D woven material using an extensive finite element mesh with 108 (27x4x1) unit structures and total degrees of freedom of 2,671,534. It is shown that the measured in-plane elastic constants correlate well with those predicted using various model, and the agreement between the experimental and predicted results are within 10% for all three in-plane constants.

Table 5.1 Comparison of predicted and measured in-plane elastic constants for 3D orthogonal woven carbon fibre reinforced composites

Model	E_1 (GPa)	E_2 (GPa)	v_{12}
Analytical Laminate block model[a]	38.39	50.88	0.034
FEA Laminate Block Model[a]	39.70	51.09	0.033
FEA Unit Cell Model[b]	40.63	49.00	0.037
Average experimental results[c]	40.97	47.30	0.035

a: Tan et al (2000b); b: Kim et al (2001); c: Tan et al (2000a)

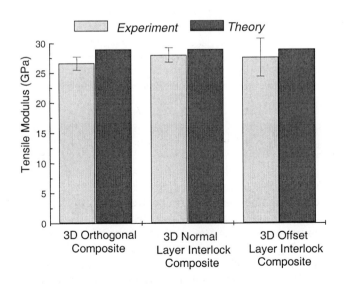

Figure 5.9 Comparison of experimental and theoretical Young's modulus values for three types of 3D woven composite.

Tan et al (2001) also measured and predicted the in-plane elastic constants for 3D orthogonal woven E-glass/epoxy composites. Table 5.2 compares the in-plane Young's moduli, the shear modulus and the Poison's ratio that were measured experimentally and predicted using both the analytical and finite element analysis based laminate block models. A good agreement between the experimental and predicted results is noted.

Table 5.2 Comparison of predicted and measured in-plane elastic constants for 3D orthogonal woven E-glass/epoxy composites

Model	E_1 (GPa)	E_2 (GPa)	v_{12}	G_{12} (GPa)
Analytical Laminate block model	29.59	27.05	0.1342	4.4790
FEA Laminate Block Model	29.46	28.03	0.1329	5.3987
Average experimental results	31.37	29.68	0.1158	4.5289

A unique feature of many 3D woven composites is that they begin to permanently deform or 'soften' at relatively low tensile stress levels (Callus et al., 1999; Ding et al., 1993; Guess and Reedy, 1985; Lee et al., 2002). This softening is shown by the kink in the stress-strain curve for the 3D composite in Figure 5.7, which does not usually occur in 2D laminates. The softening can reduce the stiffness by 20 to 50%, depending on the type of composite, and is attributed to the onset of plastic deformation of the most heavily distorted load-bearing tows, as depicted in Figure 5.4 (Cox et al., 1994; Callus et al., 1999). As reported earlier, the load-bearing tows in a 3D woven composite can be severely misaligned from the in-plane direction by the z-binders. These heavily distorted tows begin to plastically straighten when the applied tensile strain reaches a critical value sufficient to induce permanent shear flow of the resin within the fibre bundle. The critical tensile stress (σ_a) for plastic tow straightening can be estimated by (Cox et al., 1994):

$$\sigma_a = \frac{f_s |\tau_{13}|}{|\xi|} \tag{5.1}$$

where f_s is the volume fraction of load-bearing tows, $|\tau_{13}|$ is the axial shear strength of the tow, and $|\xi|$ is a fibre waviness parameter which is defined as the average misalignment angle for 90% of all load-bearing tows. Using this equation, the effect of fibre waviness on the plastic tow straightening stress is plotted in Figure 5.10. Shown in this figure are typical fibre waviness values for prepreg tape, 2D woven and 3D woven composites. From this figure it is obvious that tensile softening of 3D woven composites occurs at much lower stress values than 2D composites. Therefore, to overcome this softening it is necessary to minimise in-plane fibre waviness or use a resin having a high yield shear strength.

At tensile stresses above the onset of plastic tow straightening, 3D woven composites experience matrix cracking (both tensile and delamination), z-binder debonding, tow rupture and, in some materials, tow pull-out (Callus et al., 1999; Cox et al. 1994; Lee et al., 2000). Tensile failure generally occurs by rupture of the load-bearing tows, which may have been significantly weakened by damage incurred in the 3D weaving process (see Figure 5.3). As a result, the tensile strength of a 3D woven composite is often lower than for an equivalent 2D woven composite with a similar fibre volume content (Brandt et al., 1996; Cox and Flanagan, 1996; Lee et al., 1992). Figure 5.11 presents a compilation of published tensile strength data for 3D woven composites with different z-binder contents. In this figure the tensile strength of a 3D woven composite is normalised to the strength of the equivalent 2D laminate. It is seen that the failure strength of 3D woven composites is the same or, more often, less than the strength of the 2D laminate. It is interesting to note, however, that the tensile strength of a 3D composite is rarely more than 20% lower than the strength of the 2D material, and furthermore the tensile strength is not affected significantly by the z-binder content for the range plotted here. The lower tensile strength of 3D woven composites is due to fibre damage incurred during the weaving process that weakens the low-bearing tows (see Figure 5.3), increased fibre waviness, and pinching of the surface tows (see Figure 5.4).

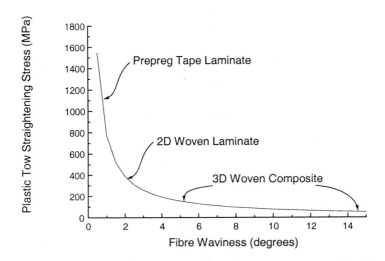

Figure 5.10 Effect of in-plane tow waviness on the tensile stress for plastic tow straightening. Representative tow straightening stresses for 2D and 3D woven composites are indicated. The comparison is made for composites with identical fibre content ($f_s = 0.3$) and shear strength ($|\tau_{13}| = 45$ MPa) values.

Predicting the tensile failure strength of 3D woven composite by micromechanical modelling is more difficult than determining the Young's modulus. This is because the extent of fibre damage, waviness and crimping are often not accurately known, and therefore it is difficult to predict the tensile stress for tow rupture. Tan et al (2000a,b; Tan et al 2001) measured and predicted the in-plane tensile strengths for both the 3D orthogonal woven CFRP (as shown in Figure 5.6) and a 3D orthogonal woven E-glass/epoxy composite.   Figure 5.12 shows micrographs of the fracture surface for specimens loaded in the stuffer yarn and filler direction, respectively. The breakage of z-binders shown in Figures 5.12(a) and (b) indicate that stuffer yarns break at a cross section between two adjacent six filler yarns. The separation of a z-binder shown in Figures 5.12(c) and (d) clearly indicates that filler yarns break at a cross section along a z-binder.

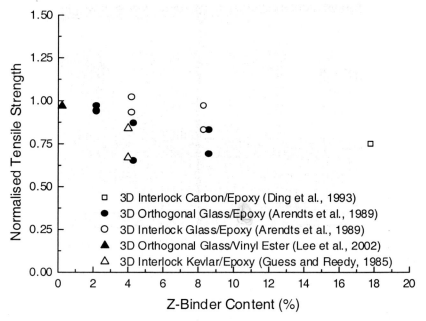

Figure 5.11 Plot of normalised tensile strength against z-binder content for various 3D woven composites.

Table 5.3 presents a comparison between the experimental and predicted in-plane tensile strengths in both the stuffer and filler yarn directions. The subscript 1 and 2 refer to the stuffer and filler yarn direction respectively.   The predicted results are obtained by using the rule of mixture method and the laminate block models in conjunction with maximum stress criterion. Both analytical method and finite element method are employed in the block laminate model.   It is noted that there exists a good correlation between the predicted and measured tensile strength in the stuffer yarn direction.   However, there is a large difference between the predicted and measured tensile strength in the filler yarn direction. This is due to the misalignment in the filler yarn direction as shown in Figure 5.6(b). Although the filler yarn is 20% more than the stuffer yarn, the average tensile strength in the filler yarn direction is only slightly larger than that in the stuffer yarn direction.   The misalignment of filler yarns is shown in Figure 5.6(b) and is believed to be the major contributing factor to the low tensile strength in the filler yarn direction.

Table 5.3 Comparison of predicted and measured in-plane tensile strengths (MPa) for 3D orthogonal woven CFRP composites

Failure strength	Rule of mixture	Laminate block model (analytical)	Laminate block model (FEA)	Exp. (avg.)
σ_{1t}	480.7	538.1	473	483.7
σ_{2t}	667.2	711.0	703	486.2

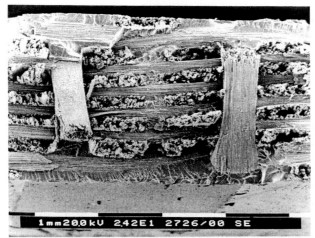

(a) Micrograph of the fracture surface for a CFRP specimen loaded in stuffer yarn direction

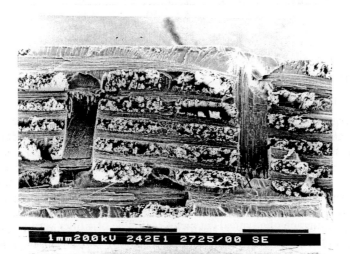

(b) Micrograph of the fracture surface opposite to that in (a)

Figure 5.12 Micrographs of the fracture cross-section for a typical CFRP specimen loaded in tension in stuffer yarn direction (Tan et al, 2000a,b)

(c) Micrograph of the fracture surface for a CFRP specimen loaded in filler direction

(d) Micrograph of the fracture surface opposite to that in (c)

Figure 5.12 (continued) Micrographs of the fracture cross-section for a typical CFRP specimen loaded in tension in filler direction (Tan et al, 2000a,b)

The influence of the misalignment can be taken into account by employing the curved beam model described in Chapter 4 (Tong et al, 2002). To employ the model, let us consider the micrograph of a typical cut along centreline of a filler yarn in the filler yarn direction as shown in Figure 5.13(a). The repeating unit of all filler yarns is marked and can be idealised as these filler yarn segments shown in Figure 5.13(b). It is further assumed that each filler yarn segment is supported by an elastic foundation and there is no interaction between filler yarn segments. The path of the centreline of each filler yarn is then measured and is idealised as a sine function with an amplitude of h_f and a half wave length of l_f. Figures 5.13(c) and (d) compare the measured and idealised paths of centreline of two filler yarn segments. Table 5.4 lists the amplitudes and half wavelengths of all filler yarn segments as shown in Figure 5.13(b). The tensile stresses at which failure occurs in all six misaligned filler yarn segments in open mode range from 483.32 to 533.07 MPa, and those in shear mode range from 437.87 to 462.21 MPa. These predicted results correlate well with the measured failure strengths in the filler yarn direction ranging from 445.1 to 509.2 MPa.

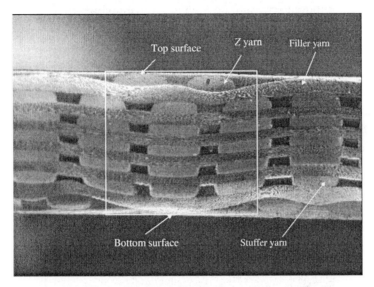

Figure 5.13a Micro-photo for a typical cross-section cut along the filler yarn direction for a 3D orthogonal CFRP composite material (Tan, 1999; Tan et al, 2000a,b)

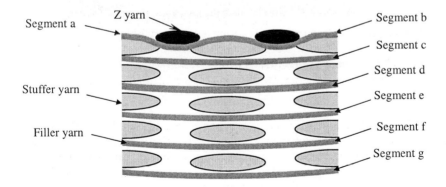

Figure 5.13b Schematic of idealised filler yarns for the 3D orthogonal CFRP composite material (Tan, 1999; Tong et al, 2002)

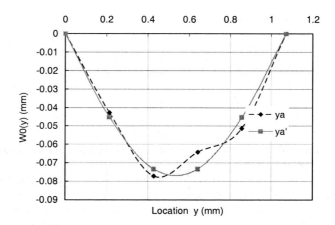

Figure 5.13c Comparison between the true waviness and ideal sine curve for the filler yarn segment a

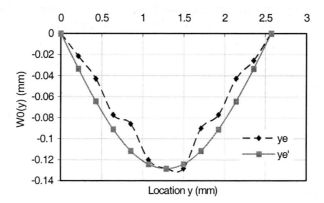

Figure 5.13d Comparison between the true waviness and ideal sine curve for the filler yarn segment e

Table 5.4 Maximum amplitudes and span lengths of the waved filler yarn segments a-g

Yarn segment	h_f (mm)	L_f (mm)
a	0.077	1.071
b	0.107	1.285
c	0.12	2.57
d	0.133	2.57
e	0.129	2.57
f	0.129	2.57
g	0.107	2.57

5.3.2 Compressive Properties

The compressive properties and failure mechanisms of 3D woven composites have been investigated in great detail because of their potential application in aerospace structures. Most attention has been given to 3D carbon/epoxy composites because of their use in aircraft, although 3D carbon/bismaleimide and 3D Kevlar/epoxy have also been examined. Most studies find that the compressive modulus of 3D woven composites is lower than 2D prepreg tape or woven laminate with a similar fibre volume content (Brandt et al., 1996; Guess and Reedy, 1986; Farley et al., 1992). The reduced modulus is due to crimping and increased waviness of the load-bearing fibres caused by the z-binders.

The effect of z-binder reinforcement on the axial compressive strength of 3D woven composites is complex, with both improvements and reductions to strength being observed. Figure 5.14 presents compressive strength data for three types of 3D composites with different z-binder contents. The normalised compressive strength is defined as the compressive strength of the 3D woven composite divided by the strength of an 2D woven laminate with nominally the same fibre content. The data plotted in Figure 5.14 shows no clear effect; with both an increase and reduction to strength occurring. The data does reveal, however, that the compressive strength of a 3D woven composites is usually improved or degraded by less than 20%, which is the same effect observed for the tensile properties shown in Figures 5.8 and 5.11.

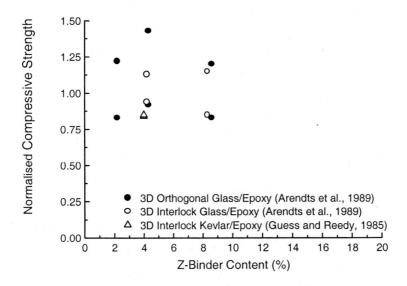

Figure 5.14 Plot of compressive strength against z-binder content for various 3D woven composites.

The cause for the improved compressive strength of the 3D woven composites is not clear. Those studies that report an improvement to the strength do not describe the compressive failure mechanisms of the 2D and 3D composites, which may shed light on the cause of the improvement. In comparison, the cause for the reduction to the compressive strength of 3D woven composites is understood due to the work of Cox et al. (1992, 1994) and others. Cox et al. (1992, 1994) and Kuo and Ko (2000) observed that 3D composites fail in axial compression by kinking of the load-bearing tows. Kinking is a failure process that initiates at regions with a low resistance to permanent shear deformation, such as at material defects (eg. void, crack) or where fibres are misaligned from the load direction. Kinking commences when the applied compression stress reaches a sufficient level to induce plastic shear flow of the resin matrix within and surrounding an axial tow. Plastic yielding of the resin allows the fibres within an individual tow to rotate in parallel. The fibres continue to rotate under increasing load until the tow becomes unstable and then breaks along a well-defined plane known as a kink band, as shown in Figure 5.15. In 2D unidirectional laminates, clusters of coplanar kink bands grow unstably which lead to sudden compression failure.

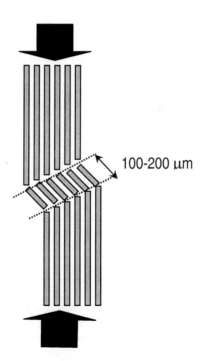

100-200 µm

Figure 5.15 Schematic of a kink band in a compressed fibre tow

The kinking failure mechanism in 3D woven composites is somewhat different to the failure event for 2D laminates. The kink bands in 3D woven composites first initiate in the most severely distorted tows, which are usually at the surface where they are pinched by the z-binders (see Figure 5.4). Cox et al. (1992) observed that two kink bands often form in the pinched tow immediately adjacent to the surface loop of the z-binder, as shown in Figure 5.16. Once the surface tow has failed it loses stiffness, but is constrained from buckling outwards by the surface loop.   Upon further loading kink bands form in other distorted tows.   Cox et al. (1992, 1994) found that kink bands within 3D woven composites develop as discrete geometric flaws rather than as coplanar bands that occur in unidirectional laminates.   As a result, 3D woven composites fail gradually at discrete locations throughout the material, leading to very high strains to ultimate failure.

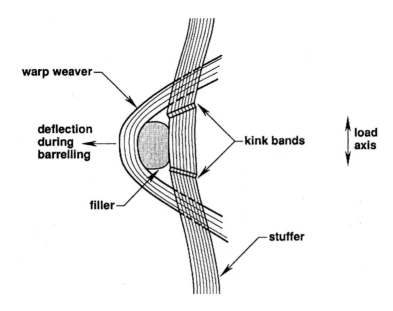

Figure 5.16 Schematic showing the locations of two kink bands within a surface axial tow (From Cox et al., 1992).

The high failure strains of 3D woven composites under compression loading is shown in Figure 5.17 that compares the compressive stress-strain behaviour of a unidirectional carbon/epoxy prepreg tape against a 3D carbon/epoxy composite measured by Cox et al. (1992).   It is seen that the curve for the tape laminate increases steadily until catastrophic failure occurs at a strain of ~1.4%. In contrast, the curve for the 3D composite shows a sudden load drop at a strain of ~0.5%, although complete failure does not occur. Instead, the load decreases very gradually over a large strain. Cox et al. (1992) found that 3D woven composites still have significant strength after compressive strains of more than 15%, indicating extraordinary high ductility. This extreme ductility is a unique property of 3D woven composites, and is due to kink bands forming as discrete geometric flaws that inhibit catastrophic failure which facilitates the gradual failure of the material under increasing strain.

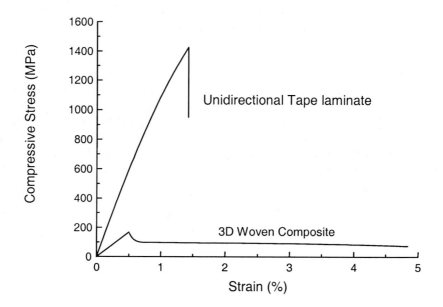

Figure 5.17 Compression stress-strain curves for a unidirectional tape laminate and a 3D woven composite (The curves for the laminate and 3D composite are from Daniel and Ishai (1994) and Cox et al. (1992), respectively).

The compressive properties of 3D woven composites under fatigue loading have also been investigated. Dadkhah et al. (1995) examined the compression-compression fatigue performance of various 3D woven carbon/epoxy composites, and found that the fatigue failure mechanism is similar to the failure process described above for monotonic compression loading. Under cyclic compression loading, a kink band initiates at a site of high fibre distortion, with the most common location being where the surface tow is crimped by the z-binder (see Figure 5.4). Upon further load cycles the fibres within the crimped tow progressively rotate to greater angles until failure occurs by kinking. It is believed that the fatigue life does not extend greatly beyond the formation of the first few kink bands. The fatigue endurance of 3D woven composites have not yet been compared against 2D laminates with the same fibre content, although it is expected to be lower due to the heavily distorted fibres lowering the cyclic stress needed to induce kinking failure.

5.3.3 Flexural Properties

The flexural properties of 3D woven composites has been investigated by Chou et al. (1992), Cox et al. (1994), Ding et al. (1993) and Guess and Reedy (1985). In most cases the flexural properties of a 3D woven composite are lower than for an equivalent 2D laminate. In the worst reported case, Guess and Reedy (1985) found that the flexural modulus and strength of a 3D Kevlar/epoxy composite was respectively 20% and 30% lower than a 2D laminate. The reduced flexural properties are due to the crimping and increased misalignment of the tows by the z-binders.

5.3.4 Interlaminar Shear Properties

The interlaminar shear strength of various types of 3D woven composite have been evaluated, and it is generally found that the strength is the same or slightly higher than an equivalent 2D material (Brandt et al., 1996; Chou et al., 1992; Ding et al., 1993; Guess and Reedy, 1985; Tanzawa et al., 1997). Figure 5.18 shows the normalised interlaminar shear strength values for four types of 3D woven composite with different z-binder contents. The normalised shear strength is the interlaminar shear strength of the 3D woven composite divided by the interlaminar strength of a 2D woven laminate with nominally the same fibre content. It is seen in Figure 5.18 that in a few cases there is an improvement to the interlaminar shear strength with the 3D woven composite, although in most cases there is no significant change.

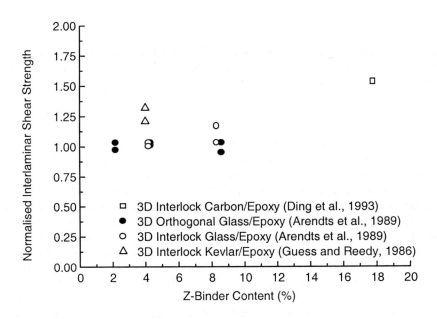

Figure 5.18 Plot of normalised interlaminar shear strength against z-binder content for various 3D woven composites.

5.4 INTERLAMINAR FRACTURE PROPERTIES OF 3D WOVEN COMPOSITES

An important advantage of 3D woven composites over conventional 2D laminates is a high resistance to delamination cracking. 2D laminates are prone to delamination cracking when subject to impact or out-of-plane loads due to their low interlaminar fracture toughness properties. 2D laminates made of thermoset prepreg material, such as carbon/epoxy tape, are particularly susceptible to delamination damage due mainly to the poor fracture toughness of the resin matrix. The low delamination resistance of carbon/epoxy is a significant factor impeding the more widespread use of these laminates in aircraft structures prone to impact from stone and bird strikes, such as the leading edges of wings and tail-sections. The superior delamination toughness of 3D woven composites has been a strong incentive for the use of these materials in highly loaded or impact prone aircraft structures such as wing panel joints (Wong, 1992), flanges, and turbine rotors (Mouritz et al., 1999).

The delamination resistance of 3D woven composites has been characterised for the mode I and II load conditions. The mode I condition is also known as tensile crack opening and mode II as shear crack sliding. Most delamination studies on 3D woven composites have been for mode I loading (Byun et al.,1989; Guenon et al., 1989; Arendts et al., 1993; Tanzawa et al., 1997; Mouritz et al., 1999). Little work has been performed on the mode II delamination properties, and this is an area requiring further research because delaminations caused by impact can propagate as shear cracks. The delamination properties of 3D woven composites subject to mode III (tearing) loading have not yet been determined, due possibly to the difficultly in performing mode III fracture tests on 3D materials.

The mode I interlaminar fracture properties of 3D composites with an orthogonal or interlocked woven structure have been thoroughly investigated (Byun et al.,1989; Guénon et al., 1989; Arendts et al., 1993; Tanzawa et al., 1997; Mouritz et al., 1999). It is found that the mode I delamination resistance of 3D woven composites is superior to 2D laminates. The delamination toughness increases with the volume content, elastic modulus, tensile strength and pull-out resistance of the z-binders. However, even relatively modest amounts of z-binder reinforcement can provide a large improvement to the delamination resistance. For example, Guénon et al. (1989) found that the delamination toughness for a 3D carbon/epoxy composite with a z-binder content of only 1% was about 14 times higher than a 2D carbon/epoxy prepreg laminate. Increasing the z-binder content can promote even larger improvements to the mode I interlaminar fracture toughness. The largest reported increase is for a 3D woven composite with an 8% binder content that has a mode I delamination resistance more than 20 times higher than for a 2D laminate (Arendts et al., 1993). Such large improvements to delamination resistance are comparable to that found with other types of 3D composites, such as knitted, stitched and z-pinned materials which will be described later.

The high mode I interlaminar fracture toughness of 3D woven composites is due to a number of toughening processes caused by the z-binders, and these are shown schematically in Figure 5.19. When a delamination starts to grow between the plies in a 3D woven composite, the crack tip passes around the z-binders without causing them any damage. In some materials the z-binders may debond from the surrounding composite when the interfacial adhesion strength is poor, although the binders themselves remain undamaged. The energy needed to debond the binders induces some toughening of the composite (Guénon et al., 1989), although it is likely to be small. Most of the toughening occurs by the formation of a z-binder bridging zone that can extend up to ~30 mm behind the crack tip. These binders that bridge the delamination are able to carry a large amount of the applied force, which reduces the stress acting on the crack tip and thereby increases the delamination resistance.

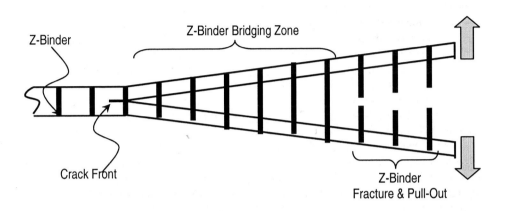

Figure 5.19 Schematic of mode I delamination cracking in a 3D woven composite.

Figure 5.20 shows a z-binder bridging a delamination in a 3D woven composite. In 3D woven composites the z-binder yarns may be distorted during manufacture (as shown in Figure 5.5), and this reduces the effective stiffness and strength of the bridging z-binders. Therefore, the interlaminar toughness provided by the z-binders is often not as high as when they are aligned normal to the crack plane. In addition, the damage incurred by the z-binder yarns during weaving reduces their tensile strength, which will also affect the interlaminar toughness. The high toughness from the z-binders often causes extensive crack branching in 3D woven composites that promotes further toughening. The applied stress acting on the binders within the bridging zone is not equal; but rather a low stress is exerted on the binders close to the crack tip and a larger stress on binders at the rear of the bridging zone where the crack opening displacement is the greatest. The binders at the rear of the bridging zone eventually break along the crack plane (as shown in Figure 5.21) or near the outer surface of the composite where the binder has been weakened during weaving in order to form a tight bend (see Figure 5.4). When the binders break near the outer surface they are gradually pulled-out of the composite, and the work done during the binder pull-out also adds to the toughness. Figure 5.22 shows a binder yarn standing proud of the fracture surface of a 3D woven composite after being pulled-out. In summary, the superior mode I delamination toughness of 3D woven composites is due to the toughening processes of debonding, bridging and pull-out of the z-binders and crack branching.

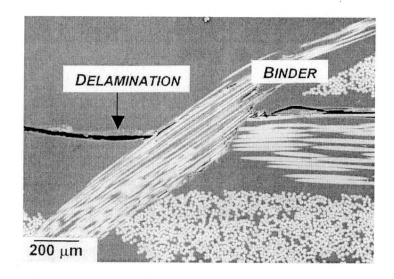

Figure 5.20 Z-binder bridging a delamination crack in a 3D woven composite (with an offset layer interlock structure) (From Mouritz et al., 1999)

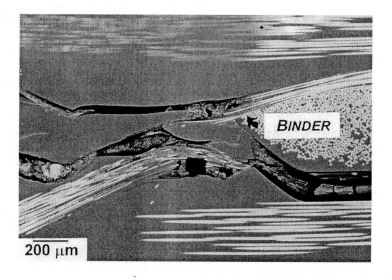

Figure 5.21 A broken z-binder beyond the rear of the bridging zone in a 3D woven composite (with an offset layer interlock structure) (From Mouritz et al., 1999).

Figure 5.22 A broken z-binder that has been pull-out out of a 3D woven composite (with an offset layer interlock structure) (From Mouritz et al., 1999).

Despite the complexity of the delamination crack growth process in 3D woven composites, it is possible to model some of the toughening processes in order to predict the mode I interlaminar fracture toughness. Byun et al. (1989) developed a finite element model that focussed on the relative contributions of matrix toughness and z-binder reinforcement to the delamination resistance. The toughening processes of binder bridging and pull-out were considered in the model, while crack branching was not included because of the difficulty in accurately modelling this process. A comparison of theoretical failure values for delamination cracking predicted using the model against experimental values for a 3D orthogonal composite is shown in Figure 5.23. The model is found to be inaccurate for short crack lengths, but the accuracy increases with the crack length. Modelling work by Byan et al. (1989) reveals that the mode I interlaminar fracture toughness of 3D woven composites can be enhanced by increasing the Young's modulus, tensile strength and volume content of the binders.

While the mode I interlaminar fracture toughness properties of 3D woven composites have been examined in detail, the mode II properties have not been thoroughly analysed. The only reported study into mode II delamination resistance was performed by Liu et al. (1989) on a 3D orthogonal carbon/epoxy composite with a small amount (1%) of z-binders. It was found the critical strain energy release rate for crack propagation (G_{IIc}) in the 3D woven composite is 2 to 3 times higher than for a 2D carbon/epoxy prepreg laminate. Liu et al. (1989) suggest that the improved mode II delamination resistance is due to debonding, bridging and shear deformation of the z-binders as well as crack branching. It is also likely that the waviness of the in-plane woven fibres also contributed to the improved toughness. It is worth noting that z-binders are more effective in improving the delamination resistance for mode I than mode II loading. This is shown in Figure 5.24 that compares improvements to the modes I and II interlaminar fracture toughness values for the same 3D woven composite. It is seen that the mode I toughness of the 3D composite is ~14 times higher than the 2D laminate whereas the mode II toughness is only 2.7 times higher.

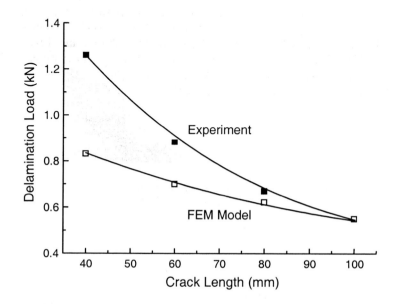

Figure 5.23 Comparison of theoretical and experimental critical load values to induce mode I delamination cracking in a 3D woven composite (Data from Byan et al., 1989).

5.5 IMPACT DAMAGE TOLERANCE OF 3D WOVEN COMPOSITES

The impact damage tolerance of 3D woven composites has been extensively evaluated because of their potential use in aircraft and rocket structures prone to impact loading. 3D woven composites have been impact tested under low to medium energy levels using light-weight, low-speed projectiles to evaluate their damage resistance for aircraft structures subject to hail and bird strikes during flight and to dropped tools during maintenance (Arendts et al., 1993; Billaut and Roussel, 1995; Brandt et al., 1996; Chou et al., 1992; Herricks and Globus, 1980; Ko and Hartmann, 1986; Reedy and Guess, 1986; Susuki and Takatoya, 1997). 3D woven composites have also been tested under ballistic impact conditions using high-velocity bullets to determine their impact damage tolerance for military aircraft and armoured vehicle applications (James and Howlett, 1997; Lundland et al., 1995).

It is found that the amount of impact damage caused to 3D woven composites is less than 2D laminates with the same fibre volume content. For example, Figure 5.25 shows the effect of increasing impact energy on the amount of delamination damage experienced by 3D carbon/epoxy composites reinforced with an orthogonal or interlocked woven structure (Billaut and Rousell, 1995). The amount of damage suffered by an aerospace-grade carbon/epoxy laminate is shown for comparison. It is seen the amount of impact damage experienced by the 3D woven composites is lower than the 2D laminate. The outstanding damage resistance of 3D woven composites is due to their high delamination resistance. The interlaminar toughening mechanisms described in the previous section, namely debonding, fracture, pull-out and, in particular, crack bridging of the z-binders impede the spread of delaminations from the impact site. The superior impact damage resistance of 3D woven composites usually results in higher post-impact mechanical properties than for 2D laminates (Arendts et al., 1993; Brandt et al., 1996; James and Howlett, 1997; Susuki and Takatoya, 1997; Voss et al., 1993). For example, in Figure 5.26 it is shown that the post-impact flexural and compressive strengths of 3D woven composites are significantly higher than for 2D laminates (Arendts et al., 1993; Voss et al., 1993).

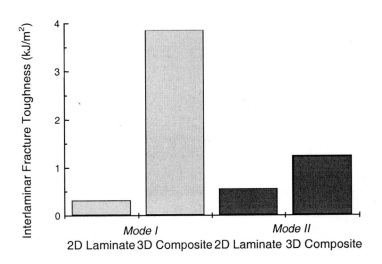

Figure 5.24 Comparison of the delamination resistance of 2D and 3D composites for mode I and II loading (The modes I and II data is from Guénon et al. (1989) and Liu et al. (1989), respectively).

5.6 3D WOVEN DISTANCE FABRIC COMPOSITES

Sandwich material made of a 3D woven fabric called 'Distance Fabric' is a relatively new class of ultra-light weight composite. 3D woven sandwich composites were originally conceived and manufactured by Verpoest and colleagues at the Katholieke Universiteit Leuven in Belgium. The composites were developed to overcome many of the shortcomings of conventional sandwich composites, which consist of two thin laminate face skins that are adhesively bonded to a light-weight core of honeycomb or rigid foam. The disadvantages of standard sandwich materials is that the manufacturing process can be labour intensive because the skins must be manufactured separately and then bonded to the core in a second processing step. Consequently, sandwich composites can be expensive to manufacture for low-cost applications such as civil and marine structures. Further problems with sandwich composites are that they are susceptible to skin-to-core separation due to bond-line defects and experience skin delamination under excessive bending, buckling or impact loads.

Advanced sandwich composites made of distance fabric offer the potential to overcome these problems. The 3D fibre architecture of a distance fabric is shown schematically in Figure 2.11, and is characterised by through-thickness fibres, known as piles, interconnecting two woven face skins. The fabric is produced using the velvet weaving process that is described in Chapter 2, and the process can be controlled to produce fabrics with different amounts and orientations of the pile yarns. After weaving, the hollow core of the fabric can be filled with a polymer or syntactic foam by liquid foam injection. The skins can be impregnated with thermoset or thermoplastic resin using the moulding processes outlined in Chapter 3.

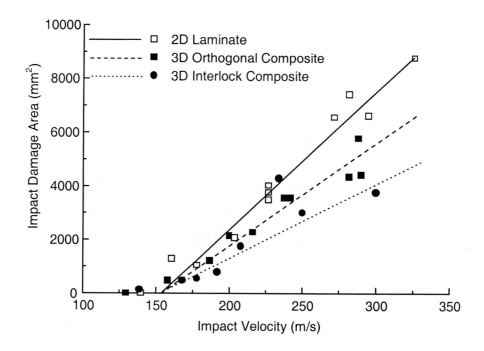

Figure 5.25 Effect of impact velocity on the amount of delamination damage to 2D and 3D woven composites (Data from Billaut and Rousell, 1995).

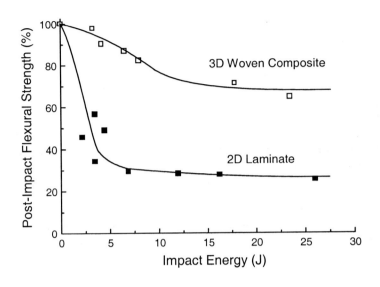

(a)

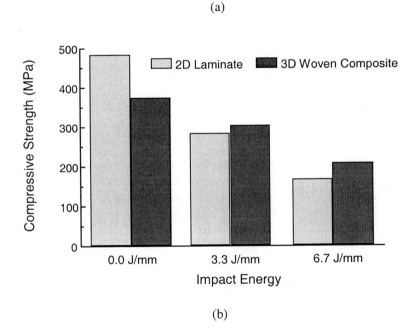

(b)

Figure 5.26 Effect of impact energy on the (a) flexural strength and (b) compressive strength of 2D and 3D woven composites. The flexural and compressive data is from Voss et al. (1993) and Arendts et al. (1993), respectively.

3D woven sandwich composite is a new class of material, and much remains unknown about its mechanical performance. Research at the Katholieke Universiteit Leuven has shown that the mechanical properties are strongly influenced by the areal density, length, angle and degree of stretching of the pile yarns, although generally properties such as compressive strength, shear strength and fatigue endurance are similar to those of conventional honeycomb materials (Judawisastra et al., 1989, Preller et al., 1990, Van Vuure et al., 1994). The key advantages of 3D woven sandwich composite is an exceptionally high delamination resistance – with a skin peel strength up to four times higher than conventional sandwich materials – and good impact damage resistance due to the integral nature of the 3D fibre structure (Preller et al., 1990, Van Vuure et al., 1994). It is reported by Verpoest and colleagues that 3D sandwich composites are used in a wide range of applications including hard-tops for cars, radar domes, mobile homes, a small aircraft, and furniture and interior wall panels for fast ferries.

Chapter 6

Braided Composite Materials

6.1 INTRODUCTION

Much of the current knowledge behind the technologies used to manufacture 3D braided preforms was generated in a period of time between the early 1980's and the late 1990's. Mostly funded through the US Government, research programs, of which the NASA Advanced Composite Technology (ACT) Program was a major focus, brought together preform suppliers such as Atlantic Research Corporation and Drexel University, with research laboratories (University of Delaware, NASA Langley, Drexel University, etc) and aerospace end-users (Boeing, Douglas and Lockheed). It was during this period that some of the more significant studies on the mechanical behaviour of 3D braided composites were performed.

However, in common with the other forms of 3D textile composites described in this book, the extent of the published literature on the mechanical properties of 3D braided composites would only constitute a small part of the information necessary to fully characterise this class of composite material. In Section 2.3 the main techniques of producing 3D braided preforms were described. Each of these manufacturing processes would result in preforms whose final consolidated properties would be influenced not only by the characteristics of the process itself but also by the variations in braid architecture that can be generated within each manufacturing technique. Figure 6.1 illustrates the highly interlinked nature of a 3D braid and critical factors such as the yarn size, the angle of the braiding yarns, the percentage content of axial yarns, etc, all have a major influence upon the resultant composite properties.

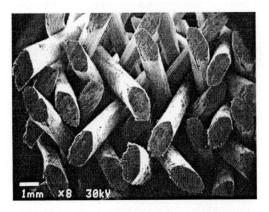

Figure 6.1 Photomicrograph of a 3D braided architecture (courtesy of Atlantic Research Corporation)

In spite of the limited data available in the published literature there are some general conclusions that can be drawn on the mechanical properties of 3D braided composites and these are summarised in the following sections.

6.2 IN-PLANE MECHANICAL PROPERTIES

Two comprehensive studies of the in-plane mechanical properties of 3D braided composites were carried out in the mid-1980's by Macander, Crane & Camponeschi (1986) and Gause & Alper (1987). In these two publications the effect of changes to a number of braiding variables on the tensile, compressive and other in-plane properties were investigated. Data was generated on preforms constructed by the 4-step, or row-and-column, braiding process.

6.2.1 Influence of Braid Pattern and Edge Condition

In the first part of their study, Macander et al. (1986) examined the effect of braid pattern and edge condition upon the performance of braids manufactured from T300 30K carbon yarns, impregnated with epoxy resin. The results of this work are summarised in Table 6.1. The braid notations used refer to the motion of the yarn carriers within the flat, cartesian plane of the 3D braider. The first number in the braid pattern designates the number of spaces the yarn carrier advances in the x-direction whilst the second number represents spaces moved in the y-direction. The use of a third category, e.g. ½F, refers to the number of carriers that remain fixed in the axial direction (½F = 50%).

Table 6.1 Braid pattern and edge condition effect on 3D braided carbon/epoxy composites (from Macander et al., 1986)

	1x1 uncut	1x1 cut	3x1 uncut	3x1 cut	1x1x ½F uncut	1x1x ½F cut
Fibre volume fraction (%)	68	68	68	68	68	68
Braiding yarn angle (°)	± 20	± 20	± 12	± 12	± 15	± 12
Tensile strength (MPa)	665.5	228.7	970.5	363.7	790.6	405.7
Tensile modulus (GPa)	97.8	50.5	126.4	76.4	117.4	82.4
Compressive strength (MPa)		179.5		226.4		385.4
Compressive modulus(GPa)		38.7		56.6		80.8
Flexural strength (MPa)	813.5	465.2	647.2	508.1	816	632.7
Flexural modulus (GPa)	77.5	34.1	85.4	54.9	86.4	60.8
Poisson's ratio	0.875	1.36	0.566	0.806	0.986	0.667

The most striking result comes from the difference in performance between specimens with cut and uncut edges. There is a 66% reduction in the tensile strength and at least a 40% reduction in tensile modulus for specimens with no axial fibres. Specimens with axial fibres suffered less reduction in their tensile properties (approximately 50% in strength and 30% in modulus) although it was still a significant drop in performance.

The flexural behaviour of the materials also showed significantly reduced performance when specimen edges were machined to produce cut fibres. This experimental data indicates the high sensitivity that 3D braided composites have to machining damage of the yarns on the surface. As each braiding yarn within the common 3D braiding processes will eventually travel to the specimen surface, any machining of this surface will result in the braiding yarns becoming non-continuous along the specimen length, with the resultant drop in performance. Due to the fixed nature of the axial fibres they will run parallel to the specimen surface and thus will not be affected by any machining. This results in their higher retained properties when compared to composites without axial fibres.

The data presented in Table 6.1 also illustrates the strong influence that the braiding pattern has upon the mechanical properties of the composite materials. The presence of axial fibres within the 1x1 architecture has produced a braid with an apparent braiding yarn angle (angle between the braiding yarn orientation and the specimen braid axis) less than that of the 1x1 architecture without axials. The orientation of the braiding yarns closer to the braid axis, which is the direction along which the testing has been performed, and the presence of the axial fibres themselves produces a composite with improved tensile, compressive and flexural properties. This improvement in composite performance due solely to a reduction in braiding yarn angle is also observed when comparing the properties of the 1x1 and 3x1 structures, in both cut and uncut edge state. A decrease of 8° in the braiding yarn angle resulted in an improvement in tensile and compressive performance of 25 - 50%. Wenning et al. (1993) also observed a similar improvement in the tensile modulus with a decrease in the fibre angle of 4-step braided composites.

Other investigations on the influence of braid angle were conducted by Brookstein et al. (1993), who investigated the properties of carbon/epoxy 3D composites that were braided by the Multilayer Interlock method. Specimens with two braiding patterns (i.e. differing braid angles) were tested, $\pm45°/0°/\pm45°$ ($V_f = 43\%$) and $\pm60°/0°/\pm60°$ ($V_f = 45\%$) and the results of these tests were normalised to a nominal 50% fibre volume fraction for comparative purposes (see Table 6.2). When comparing the properties of the two 3D braided patterns, Brookstein et al. also found that the tensile and compressive properties in the longitudinal direction were improved when the braiding yarn angle was reduced, but at the sacrifice of the transverse performance. The design of 3D braided preforms must therefore be a compromise between the required mechanical performance and the number of axial yarns and the braid angle possible within a certain braiding technique.

The influence of axial fibres on the composite mechanical performance was also noted by Gause et al. (1987) who observed significant increases in the longitudinal tensile and compressive properties of carbon/epoxy, 4-step braided specimens when half of the yarns available for braiding were fixed as axial yarns. Table 6.3 summarises the results of this work although it should be noted that errors in some of the data contained in the original publication were corrected in a later publication by Ko (1989) and it is from this publication that the data in Table 6.3 is taken. It is interesting to note that although the presence of axial yarns has improved the longitudinal properties of the braided specimens, it comes at the sacrifice of the transverse properties. This is because there are now fewer yarns available as braiding yarns and thus less reinforcement oriented towards the transverse direction.

yarns. Note that the strength and stiffness results for the Celion 6K specimens have also been normalised to a fibre volume fraction of 68% to allow for better comparison.

Table 6.4 Comparison of E-glass/epoxy 2-step and 4-step braided composite (from Byun et al., 1991a)

	2-step ($V_f = 40\%$)	4-step ($V_f = 40\%$)
Number of axial yarns	38	0
Number of braider yarns	11	34
Surface braider yarn angle (°)	55	25
Tensile modulus (GPa)	35.2	25.5
Tensile Strength (MPa)	502	420
Poisson's ratio	0.31	0.58
Tensile failure strain(%)	1.33	1.83
Compressive modulus (GPa)	23.1	15.3
Compressive strength (MPa)	418	194
Compressive failure strain (%)	1.87	1.4
Short beam shear strength (MPa)	71	76.8

Table 6.5 Effect of yarn size upon the mechanical properties of carbon/epoxy 1x1 4-step braided composites (from Macander et al., 1986)

	AS4 3K	AS4 6K	AS4 12K	Celion 6K	Celion 12K
Fibre volume fraction (%)	68	68	68	56 (68 normalised)	68
Tensile strength (MPa)	736.8	841.4	1067.2	857.7 (1041.5)	1219.8
Tensile modulus (GPa)	83.5	119.3	114.7	87.8 (106.6)	113.1
Short beam shear strength (MPa)	114.8	126	121.4	71.4 (86.7)	71.4
Poisson's ratio	0.945	1.051	0.98	0.968	0.874
Flexural strength (MPa)	885.3	739.8	1063.3	-	-
Flexural modulus (GPa)	84.5	95.2	136.5	-	-
Apparent fibre angle (deg.)	±19	±15	±15	±15	±17.5

The results in Table 6.5 suggest that the tensile strength and modulus of the 4-step braided composites increase with increasing yarn size. This could be related more to the dependence of the braid angle on the yarn size, as larger yarn sizes were generally observed to produce lower braiding angles in the specimens. However, Macander et al. (1986) did not propose this as the only influence that yarn size had upon the tensile properties and concluded that other variables not clearly identified play a significant role in this effect. They suggested that the effect of "crowding" of the braider yarns at the specimen edges can orient them more along the axial direction and thus improve the overall composite properties. They postulated that the extent of crowding may be more pronounced as the yarn size increases, therefore improving the tensile properties for larger yarn size specimens, however no supporting evidence was presented for this theory.

The flexural and shear properties do not show any clear trend of improvement with decreased braid angle, again indicating that the effect of yarn size, although significant, is not a clearly understood phenomenon.

6.2.4 Comparison with 2D Laminates

Gause et al. (1987) compared the properties of their 1x1 and 1x1x½F 3D braided specimens with a 24 ply laminate of AS1/3501 prepreg with a lay-up orientation designed to mimic the proportions of fibres contained in the 1x1x½F 3D braided material (Table 6.3). The authors found that there was no clear trend in the comparison of undamaged in-plane properties between 2D and 3D materials. The tensile strengths in both directions as well as transverse tensile modulus was found to be worse for the 3D braid but the longitudinal compressive properties and tensile modulus were found to be better. In the case of open hole properties the 3D braided materials retained a far greater proportion of their tensile strength than the 2D laminate, at least 86% gross strength compared to approximately 50% for the 2D laminate. However the comparison of open hole compressive strength did not follow a similar trend, although this may be due to a lower than expected value of compression strength for the 2D laminate.

It should be expected that 3D braided composites will not have undamaged, in-plane properties that match, or are superior to, 2D prepreg tape laminates of similar fibre orientation and volume fractions. This is due to the fact that the yarns within the braid will suffer from a certain level of crimping as a result of the braiding process and this will reduce their performance relative to the uncrimped fibres in the prepreg tape. A better comparison to make is of 2D and 3D braided composites and this was done within the work published by Brookstein et al. (1993). The results that are summarised in Table 6.2 also give a comparison between the properties of 2D triaxial braids and 3D Multilayer Interlock braids manufactured from the same 12K AS4 carbon tow and epoxy resin (results normalised to 50% fibre volume fraction). Except for the case of the compressive strength the results show that for both braid patterns the 2D braids have better performance in the longitudinal direction than the 3D braids but lower in the transverse direction. The authors suggested that it was possible that the 0° fibres in the 3D braids were pushed away from the axis by the geometrical configuration of the interlocking braiding yarns and therefore were improving the transverse performance of the specimens at the detriment of the longitudinal.

It is clear from the published literature that more data is needed before a strict comparison can be made between the in-plane properties of 3D braided composites and the standard 2D laminates.

6.3 FRACTURE TOUGHNESS AND DAMAGE PERFORMANCE

As with all 3D textile composites, the addition of the third dimension of reinforcement is expected to invest composites made from 3D braided material with improved toughness and damage characteristics. There has been very little published that compares the fracture toughness of 3D braided composites with other forms of composite reinforcement, therefore it is not possible at this time to make any strict comparisons as to any potential improvements. However, the mode I fracture behaviour of a 4-step braided carbon/epoxy material was examined by Filatovs et al. (1994) in a compact tension arrangement and the effect of the notch orientation relative to the direction of the braiding axis was investigated. It was found that the force required to initiate and grow a crack through the 3D braid increased by a factor of 4 as the braid axis orientation varied from in line with the notch to transverse to it. The lowest value for crack propagation force was observed when the notch axis was at the same angle to the braid axis as the braiding yarns themselves, thus allowing crack propagation to occur partially along sections of the braiding yarns. Unfortunately, the authors did not translate these results into measurements of fracture toughness and did not compare them with measurements on conventional 2D laminates.

There is more published work that examines the general damage tolerance of 3D braided composites. In their work on the general mechanical properties of 4-step braided, carbon/epoxy composites Gause et al. (1987) also compared the OHT and OHC strength of 1x1 and 1x1x½F braids with a 2D laminate (Table 6.3). The 3D braids were observed to retain a very high proportion of their undamaged gross tensile strength (99% and 86% for the 1x1 and 1x1x½F respectively) compared to the 49% retained by the 2D laminate. In compression their retained strengths were of a similar order (42-47%). Under drop weight impact tests the 3D braids were found to perform far better at limiting the extent of damage, having less than half of the damage area created at the higher test energies than the 2D laminate.

Ko et al. (1991) also examined the strength retention of carbon/PEEK 3D braids compared to 2D laminates under OHT conditions. Although the 2D laminate had far superior undamaged tensile properties (1081 MPa versus 586 MPa), it was found that with similar proportions of fibres in the 0° and ±20° directions the 3D braided specimens had a far greater retention of tensile strength than the 2D laminates (79% and 58% respectively). Impact tests were also conducted upon the specimens and it was found that the 3D braided materials had higher compression after impact strength and an order of magnitude lower damage area than the 2D laminates.

Brookstein et al. (1993) compared the CAI performance of 2D and 3D braided composites (Table 6.2) and found that at the two impact energy levels tested (3 and 7 J/mm) the 3D braided composites had approximately the same or better compression strength compared to the 2D braided samples. This less significant difference between the impact performance of 3D and 2D braids compared to 3D braids and 2D tape laminates can be understood through the general architecture of braids. Even with an absence of through-thickness braiding yarns, the architecture of a 2D braided laminate is very undulated with the layers of braided fabric nesting significantly with each-other. This makes it very difficult for impact damage to propagate extensively within the composite as compared to the relatively straight crack paths available in tape laminates.

Overall, the damage resistance and tolerance of 3D braided composites is seen to be significantly greater than that of 2D tape laminates and at least the same as, or greater than, that of 2D braided composites. However, no data exists for 3D braided polymer matrix composites that examines the effect that the braid architecture or the braiding process has upon their fracture or damage performance. Much of this investigation has been conducted in ceramic and metal matrix 3D braided composites.

6.4 FATIGUE PERFORMANCE

In their comprehensive investigation of the mechanical properties of two, 4-step braided composites, Gause et al. (1987) measured the fatigue performance of the 3D braided materials in tension-tension (T-T), tension-compression (T-C) and compression-compression (C-C) loading and compared it to a baseline 24 ply tape laminate. The data was highly scattered but at tests running to a million cycles it was clear that the baseline laminates had significantly better fatigue performance than the 3D braids. The maximum (averaged) fatigue stress, as a percentage of their ultimate static strength, that was carried successfully to one million cycles by the tape laminate specimens was found to be 73% (T-T), 50% (T-C) and 78% (C-C). This is compared to 57% (T-T), 37% (T-C) and 43% (C-C) for the 1x1 braid, and 56% (T-T), 37% (T-C) and 52% (C-C) for the 1x1x½F braid architecture. The improved fatigue performance of the 2D laminates over the 3D braids was attributed to the fibre waviness that is intrinsic to the braided architecture. This waviness allows the fibres to bend in addition to deforming axially under load, thus working the matrix more severely. In T-T and T-C fatigue conditions both braided architectures behaved identically. In C-C conditions the authors stated that the 1x1x½F braid architecture showed greater life capability then the 1x1 architecture, which they credited to the presence of the fixed 0° yarns providing greater resistance to catastrophic fatigue damage. However, the scatter in results that is evident from the published data makes it unrealistic to draw this conclusion.

Similar fatigue results were seen by Gethers et al. (1994) in their tension-tension testing of 4-step braided carbon/epoxy materials. Although the behaviour of the 3D braids was not compared to 2D laminates, the average maximum fatigue stress at one million cycles was approximately 55% of the 3D braids static tensile strength, very similar to that recorded by Gause et al. (1987). Those specimens that survived one million cycles of testing were tested to failure statically and found to have a residual tensile strength that was 80% of the original tensile strength.

6.5 MODELLING OF BRAIDED COMPOSITES

There have been a number of models developed to predict the mechanical properties of 3D braided composites and, in a similar fashion to the other 3D textile composites described in this book, these models first depend upon an accurate description of the 3D braided yarn to be made. This description is accomplished through a geometric modelling of the yarn topology that is based purely upon the braiding procedure itself. Each particular braiding process has specific, characteristic equations that govern the topology of the yarn structure within the preform. These characteristic equations are explained in greater detail for 4-step braiding by Wang et al. (1994) and for 2-step braiding by Byun et al. (1991b).

Once the geometric model of the 3D braid has been established the process of modelling its mechanical properties is carried out in a similar fashion to other 3D textile composites. A Representative Volume Element (RVE) of the braid is identified and the properties of this RVE are established through application of analysis techniques such as classical lamination theory (Byun et al., 1991b) or an elastic strain energy approach (Ma et al., 1986). The classical lamination theory was also used by Yang et al. (1986) in the development of their Fibre Inclination Model. The properties of the overall composite are then obtained by averaging the properties of the RVE for the different yarn orientations present in the global coordinate system. More recently Chen et al. (1999) described the use of a Finite Multiphase Element method to predict the elastic properties of 3D braids. This process uses a two step numerical approach of generating a fine local mesh at the unit cell level to analyse the stress/strain response of the unit cell, then a coarse global mesh to obtain the overall response of the composite at the macroscopic level.

To date these models have only been used for the prediction of elastic constants and there does not appear to be any attempt made to predict the strength of 3D braided, polymer matrix composites. The comparison of predicted and experimental elastic constants is reasonable good, mostly within 10% (Chen et al., 1999) but in general the predicted properties are less than that measured via experiment.

6.6 SUMMARY

3D braided preforms are very versatile forms of textile reinforcement for composite structures. As discussed in Chapter 2, 3D braids can be produced in a wide range of cross-sectional shapes and these shapes can be varied along their length to form structural details such as tapers, bifurcations, holes, etc. However, there has only been relatively limited data published on the mechanical properties of 3D braided polymer matrix composites, much of the development in the area of 3D braids appears to be focussed on ceramic and metal matrix composites. In particular, there has been little comparison made between the performance of 3D braided composites made by different braiding processes and between 3D braids and 2D laminates.

From the data that has been published it is evident that the presence (or absence) of axial fibres and the angle of the braiding yarns both play an important role in controlling the mechanical properties. Improved longitudinal performance results from increased axial fibre content and decreased braiding angle, but at the expense of transverse properties. The damage resistance and tolerance of 3D braided composites are also significantly better than 2D tape laminates due to the highly interlinked nature of the 3D architecture, however the fatigue performance has been shown to be worse.

A result of particular interest is the high sensitivity that 3D braided composites have to cut edges. The act of machining the specimen edge and thus cutting the braiding yarns into discontinuous sections was found to significantly decrease the tensile and flexural properties of the composite. This indicates the need to produce 3D braided composite components to net-shape, thus removing any need for machining that will reduce its performance.

Before 3D braided preforms can be generally accepted as reinforcements for composite structures, a great deal more information must be gathered on their mechanical properties. In particular, the effect on the mechanical performance of the braiding technique and the various processing parameters within each technique must be understood in order for these reinforcement styles to be used with confidence.

Chapter 7

Knitted Composite Materials

7.1 INTRODUCTION

Knitted preforms for composite reinforcement are the least understood of the four major classes of 3D fibre preforms constructed through textile manufacturing processes. Knitted preforms are also regarded by many as not true three-dimensional reinforcements. While it is true that much of the research conducted into knitted composites has been performed upon specimens manufactured by the lay-up of individual knitted fabric layers, current commercial knitting machines are capable of producing fabric containing up to four interconnected knit layers. Most conventional "two-dimensional" knitted fabrics also contain a significant proportion of their yarns in the thickness direction of the fabric, as shown in Figure 7.1. The open nature of the knit architecture also allows a high degree of "nesting" or mechanical interlocking between individual layers of knitted fabric. These two aspects of the knit fabric architecture result in properties such as Mode I fracture toughness (outlined in Section 7.3) being significantly higher than that observed in traditional 2D woven composites.

The knitting process, which has been described in greater detail in Section 2.4, is also capable of manufacturing complex, net-shape preforms. Thus, although knitted preforms are not yet capable of being produced with similar thickness dimensions to 3D woven or braided preforms, they can be credibly included as a class of 3D textile reinforcements.

As shown in Figure 7.1 the primary difference between knitted fabrics and woven or braided, is the highly curved nature of the yarn architecture. This architecture results in a fabric that will theoretically provide less structural strength to a composite (compared to woven and braided fabrics) but is highly conformable and thus ideally suited to manufacture relatively non-structural components of complex shape. In spite of its potential markets, knowledge of the structural performance of a knitted reinforcement is still of importance if it is to be used in a composite component. However, there are inherent aspects of the knitting process which make the establishment of mechanical properties very complex. The knitting process is capable of producing a wide variety of knit architectures and within each architecture the size and shape of the loops can be adjusted to quite dramatically change the proportion of yarn length that makes up each segment of the loop (see Figure 7.2). Knitted fabric can also "relax" (i.e. yarns seeking to move to their lowest energy configuration) to a far greater degree than woven and braided fabrics. This can also result in an internal rearrangement of the knit architecture that can significantly vary the knit loop parameters in the fabric from those set on the knitting machine during the manufacturing process. When comparing fabric produced from different machines, particularly the older knitting machines, even those of the same gauge (knitting needle density) can produce the same fabric style with significantly different loop parameters, which will result in varying mechanical properties. In the more advanced, computer controlled machines this is less of an issue as the equipment is capable of controlling the loop parameters to a far greater degree.

These variations of the knit architecture can cause a broad spectrum in the mechanical performance of knitted composites and thus make it difficult to adequately compare these properties with those of other reinforcement styles. In spite of this difficulty, work has been undertaken by a number of groups, primarily since the late 1980's, to investigate and understand the performance of composites reinforced by knitted fabrics. A useful review of the area of knitted composites is given by Leong et al. (2000).

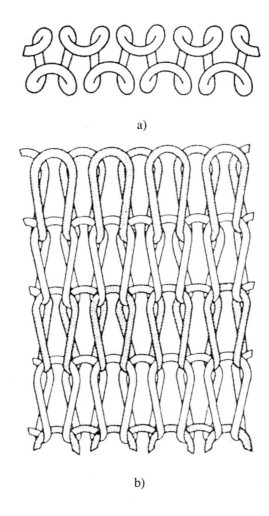

a)

b)

Figure 7.1 Illustration of the 3D nature of typical knitted fabric a) top edge view, b) plane view

Needle loop   Sinker loop   Sides or legs

Figure 7.2 Illustration of the main segments in a knit loop

7.2 IN-PLANE MECHANICAL PROPERTIES

7.2.1 Tensile Properties

The tensile performance of knitted composites has been a primary area of investigation within the published literature. Most of the investigation has focussed upon the tensile properties of weft knitted E-glass fabrics, generally with a plain knit architecture, that have been consolidated with epoxy resins. Often there is little information in the literature on the knitting parameters, such as loop lengths, shapes and densities, that will allow a more detailed analysis, however the available results allow some general behaviour to be established.

Experimental data on the tensile performance of plain, weft knitted E-glass/epoxy composites has been collated from Ramakrishna et al. (1997), Huang et al. (2001) and Höhfeld et al. (1994) and the variation of tensile strength and modulus with fibre volume fraction is illustrated in Figures 7.3 and 7.4 respectively. It can be clearly seen that the tensile performance of the plain knit E-glass/epoxy composites increases with an increase in the volume fraction of glass fibres. It can also be seen that the tensile properties are similar to that expected from randomly orientated E-glass mat reinforcements of the same fibre volume fraction. This is to be expected as the architecture of the plain knit has very few straight sections of yarn and certainly none that are of any significant length. This aspect of the plain knit architecture, as well as many other standard knit styles, will generally limited the mechanical performance of knitted composites to values much lower than that expected from conventional 2D woven fabrics (strengths of E-glass fabrics ~ 350 – 400 MPa).

Leong et al. (2000) and Anwar et al. (1997), presented tensile results of composites manufactured from other weft knit architectures (fibre volume fractions of ~ 53%). Rib (1x1) specimens had a tensile strength of 96 MPa and a modulus of 14.7 GPa, whilst full milano architectures gave strength and modulus values of 122 MPa and 14.9 GPa respectively. Clearly the style of knit architecture is influencing the tensile performance of the knitted composite. This effect of the knit architecture upon the mechanical properties of the composite is also observed in the results of Wu et al. (1993) and

Huysmans et al. (1996), both of whom investigated the properties of various warp knit architectures. Tables 7.1 and 7.2 summarise these results. In both sets of results it can be clearly seen that not only can the knitting process produce reinforcement fabrics with a broad range of properties but that a variation in a knit architecture can change a fabric from one with approximately isotropic properties to one with strongly anisotropic behaviour. In a similar fashion to woven fabrics, both the stiffness and strength of knitted composites can be improved not only by increased fibre volume fraction, but also by preferential fibre orientations within the fabric. This is illustrated in Figure 7.5, which shows the knit architectures of single dembigh, 1x3 single cord and 1x4 single cord that were examined by Wu et al. (1993). As the proportion of fibres orientated in the course direction increases so to does the tensile performance of the composite material in the course direction whilst the wale direction performance remains relatively unchanged. It should be noted that this preferential orientation of the fibres can also lead to the directional properties of knitted composites being far superior to that of random mats although still less than typical 2D woven fabrics.

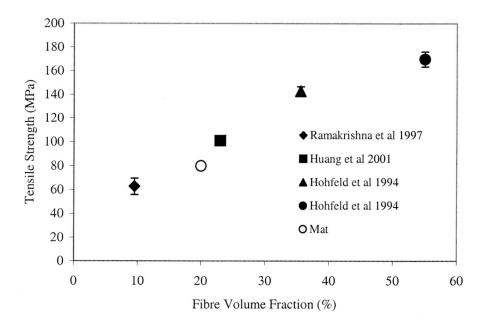

Figure 7.3 Variation in tensile strength of E-glass/epoxy, plain knit composites with fibre volume fraction

The anisotropy in the tensile performance of knitted composites has also been examined by Ha et al. (1993) and Verpoest et al. (1992) who examined the behaviour of carbon fibre (AS4)/PEEK plain knit and E-glass/epoxy plain knit composites respectively. In a similar fashion to Huysmans et al. (1996), both found that knitted architectures could be produced that had either strongly anisotropic or relatively isotropic tensile properties. This is in direct contrast with normal 2D weaves whose performance will always show a pronounced reduction in the bias direction. Verpoest et al. (1992) also noticed an improvement in tensile strength and failure strain with an increasing number of knitted fabric layers in the composite, even though the fibre volume fraction remained constant. An improvement in tensile strength with knit fabric layers was also observed by Ramakrishna et al. (1994) in 1x1 rib, weft knit carbon fibre (AS4)/epoxy specimens. This improvement in performance is attributed to a mechanical interlocking between neighbouring fabric layers, which results from the ability of knitted fabric layers to nest significantly with each other.

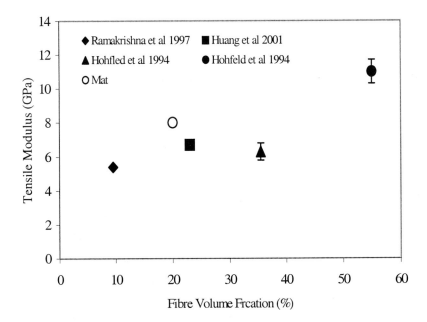

Figure 7.4 Variation in tensile modulus of E-glass/epoxy, plain knit composites with fibre volume fraction

Table 7.1 Tensile properties of Aramid/polyester warp knit composites with varying knit architecture. Fibre volume fraction constant but not reported (from Wu et al., 1993)

		Single dembigh	1 x 3 single cord	1 x 4 single cord
Strength (MPa)	wale	50.2	55	43.7
	course	80.4	154.8	187.2
Stiffness (GPa)	wale	3.44	3.69	3.41
	course	3.93	5.16	5.48

Table 7.2 Tensile properties of E-glass/epoxy warp knit composites with varying knit architecture. Fibre volume fraction = 40% (from Huysmans et al., 1996)

		Tricot	Tissue 1	Tissue 2	Tissue 3	Satin
Strength (MPa)	wale	100	100	120	100	230
	course	195	120	230	250	170
Stiffness (GPa)	wale	10	11	11	10	16
	course	13	10	15	15	13

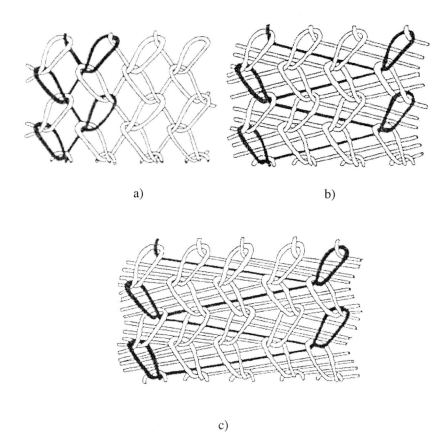

a)

b)

c)

Figure 7.5 Illustration of warp knit a) denbigh, b) 1x3 single cord, and c) 1x4 single cord architectures

The tensile properties of knitted composites can also be affected by controlling parameters, such as loop length or stitch density, within a knit architecture. Loop length and stitch density are inter-related as an increase in the number of stitches (or knit loops) per unit area will require a decrease in the length of the knit loop. Leong et al. (2000) and Anwar et al. (1997) presented data that described the effect of varying loop length and stitch density on the tensile performance of E-glass/vinyl ester weft knitted composites. The authors found that the tensile modulus in the wale or course directions is not significantly affected by varying knit parameters, being primarily controlled by the fibre volume fraction when the style of the knit architecture is unchanged. The tensile strength and failure strain in both the wale and course directions were found to decrease with a decrease in loop length (or an increase in stitch density). This behaviour is illustrated in Figure 7.6, which shows the variation in tensile strength for 3 styles of weft knit composites. This effect of stitch density upon tensile strength is contrary to that reported by Wu et al. (1993) who observed a significant increase in the course-direction tensile strength and stiffness of warp knitted aramid/polyester specimens with increasing stitch density, although no significant change was observed in the wale direction properties. Clearly there are a number of factors related to the relative proportions of the knit loop that affect its tensile performance but not enough data yet exists to provide a clear understanding.

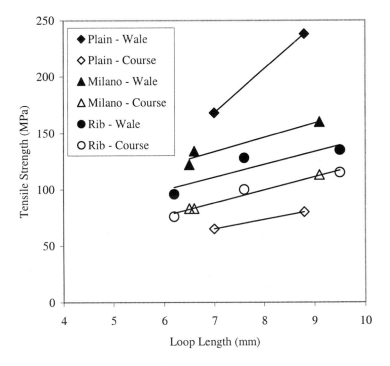

Figure 7.6 Tensile strength versus loop length for various E-glass/vinyl ester, weft knitted composites (from Leong et al., 2000)

Deformation of the knit architecture prior to consolidation with resin has also been observed to affect the tensile properties of the resultant composite material. Ha et al. (1993), Verpoest et al. (1992), Leong et al. (1999) and Khondker et al. (2001a) have all reported an improvement in tensile properties in the direction of fabric stretch. This can be clearly understood in that during fabric stretch the fibres, whose orientations were rather randomly distributed in all directions, now begin to align more towards the axis of the fabric stretch. This increase in the proportion of the fibres oriented in the direction of loading will naturally improve the tensile performance of the material.

The ability of knitted fabrics to be deformed easily and of the knitting process itself to produce holes integrally formed within the fabric, allows for the possibility of producing composites with continuous fibres surrounding a notch or bolthole rather than the broken fibres produced during the drilling of holes in composites. The effect of formed holes upon the notched tensile strength and bearing performance of knitted composites was examined by de Haan et al (1997) and Leong et al (1998) respectively. In both investigations the performance of specimens with holes formed into the knit architecture was significantly improved compared to the specimens with drilled holes (see Table 7.3). This was due not only to the unbroken yarns surrounding the hole but also to the increase in the fibre volume fraction around the hole that occurs when the hole is formed into the knitted fabric.

Table 7.3 Notched (de Haan et al., 1997) and Bearing (Leong et al., 1998) wale direction tensile properties of weft knitted composites (W/D = 4)

Structural form	Notched		Bearing	
Materials	Aramid/epoxy plain knit		E-Glass/epoxy milano knit	
Techniques	Formed	Drilled	Formed	Drilled
Strength (MPa)	75	58	338	275

The failure process of a knitted composite is, like its architecture, a complex situation. A number of researchers (Rudd et al., 1990; Ramakrishna et al., 1994; Wu et al., 1993; Ramakrishna et al., 1997; Leong et al., 1999; and Huysmans et al., 2001) have examined the various stages of tensile failure in warp and weft knitted composites, ranging from low fibre volume fraction, single layer materials, to high fibre volume fraction, multilayer specimens. It is generally accepted that the first stage of failure occurs at relatively low strain values and is the result of debonding between the resin and the portions of the knit loops orientated transverse to the loading direction (see Figure 7.7). Upon increasing load these cracks propagate into the resin-rich regions between the yarn loops. As these cracks grow and coalesce they are bridged by the unbroken yarns of the knit loops. The composite behaviour following this is then largely dependent upon the number and geometry of the yarns crossing the crack plane. Architectures with highly orientated yarns will pick up the load almost immediately whilst those with significant curvature, or off-axis orientation, may rotate or stretch before becoming fully loaded. Final failure of the knit loops has been seen to occur in either one of the two places (and often a mixture of both), at the "legs" of the knit loop where the local fibre volume fraction is lowest, or at the loop crossover points where the stress concentrations are highest.

7.2.2 Compressive Properties

Unlike the tensile properties, relatively little has been reported on the compressive properties of knitted composites. A number of researchers (Wang et al., 1995a; Leong et al., 1998; Leong et al., 1999; and Khondker et al., 2001a) have reported that for 1x1 rib and milano weft knitted composite materials, the compressive strength is significantly better than the tensile strength whilst the compressive modulus is similar to, or slightly less than, the tensile modulus. Examples of this behaviour are given in Table 7.4. Again distinctly different from the tension properties, the effect of changing loop lengths and stitch densities upon the compressive properties is far less significant. Khondker *et al* 2001b reported that in E-glass/vinyl ester weft knitted composites (plain, rib and milano architectures) at best only marginal improvement was observed in the compression strength with an increase in loop length whilst negligible effect was observed on the compressive modulus.

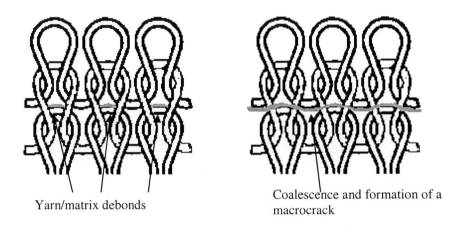

Yarn/matrix debonds

Coalescence and formation of a macrocrack

Figure 7.7 Illustration of the failure process within knitted composites

Table 7.4 Compressive and tensile properties of E-glass weft knitted 1x1 rib (from Wang et al., 1995) and milano (from Khondker et al., 2001) composites

Properties	Directions	1x1 Rib Epoxy matrix $V_f = 48\%$	Milano Vinyl ester $V_f = 55\%$
Compression Strength (MPa)	Wale	147	152
	Course	149	158
Compression Modulus (GPa)	Wale	10.8	12.5
	Course	11.1	11.8
Tensile Strength (MPa)	Wale	64.3	108
	Course	80.5	81
Tensile Modulus (GPa)	Wale	12.4	12.5
	Course	14.1	11.8

It has also been found by many of the authors that the directional properties of the knitted composites in compression are far more isotropic than the tensile properties. Leong et al. (1999) and Khondker et al. (2001a) have observed that, unlike the tensile properties, the compressive performance of the E-glass weft knit milano material is not significantly effected by any stretching of the knitted fabric prior to consolidation. Both the isotropic nature and insensitivity to stretch and knit architectural changes of the compressive properties are believed to be due to the mode of failure that occurs under compressive loading. Khondker et al. (2001a) identified that failure involved the formation of yarn kinks, which were a direct result of the buckling of the most highly curved sections of the loaded yarns (see Figure 7.8). This type of failure is very dependent upon the properties of the matrix thus, although the fibre volume fraction and knit architecture will have some effect on the compression performance, the matrix properties will tend to be the dominating factor.

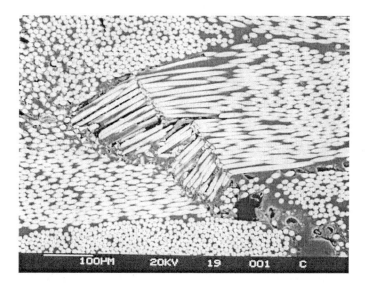

Figure 7.8 Compressive failure in knitted composites through kink formation (courtesy of the Cooperative Research Centre for Advanced Composite Structures Ltd)

7.2.3 In-Plane Properties of Non-Crimp Fabrics

In Chapter 2 the production of a specialised sub-group of knitted fabrics, known as Non-crimp fabrics, was described. Non-crimp fabrics were designed primarily as an alternative reinforcement material to tape and woven prepregs. The use of these fabrics will result in reduced costs in the lay up of composite structures due to their multilayer structure. Although produced using warp knitting techniques this family of fabrics contains substantial proportions of relatively straight, in plane yarns which will dominate the mechanical performance of the composite materials for which it is a reinforcement. Thus the proportion and orientation of these in-plane yarns will be the controlling factor in much of the mechanical performance rather than the structure of the warp knit yarns. The improvement that is observed in the mechanical performance of non-crimps over conventional knitted composites through the inclusion of straight, inlayed yarns is however achieved at the cost of the high formability of the conventional knitted fabric.

The use of non-crimp fabrics is now commonplace within the maritime industry and in the manufacture of wind turbine blades and, as pointed out in Chapter 2, it is a prime material candidate for future aircraft programs. There has been a great deal of development of this fabric style, with improvements in the visual quality of the fabric and the range of lay up options available as well as improvements in its mechanical performance.

The properties of composites reinforced by non-crimp fabrics have been examined by a number of researchers (Hogg et al., 1993; Wang et al., 1995b; Dexter et al., 1996; Bibo et al., 1997; Bibo et al., 1998). In general non-crimp composites have tension, compression and flexure properties that are inferior to laminates of similar lay up manufactured from unidirectional prepreg tape, as shown in Table 7.5. However, interlaminar shear strength is observed to improve. The variation in properties is due to the fact that the in-plane yarns within the non-crimp fabric are not completely straight. During the warp-knitting process, out of plane crimping can occur in these yarns which will degrade the resultant composite performance relative to the non-crimped prepreg laminates. By the same reasoning, in comparison to laminates manufactured from woven prepreg, non-crimp composites can exhibit superior tension, compression and flexure properties if the yarns within the woven prepreg are more undulated.

Table 7.5 Tensile and Compressive properties of Non-crimp and Unidirectional prepreg tape composites (from Bibo et al., 1997)

Properties	Non-crimp [{45,-45,0},{0,-45,45}]$_s$		Unidirectional Prepreg [45$_2$,-45$_2$,0$_6$,-45$_2$,45$_2$]$_s$	
	0°	90°	0°	90°
Tensile Strength (MPa)	621	159	951	123
Tensile Modulus (GPa)	60.8	17.2	64.8	21.4
Compressive Strength (MPa)	574	236	852	215
Compressive Modulus (GPa)	54.7	16.5	59.9	19.6
Flexure Strength (MPa)	990	310	1140	280
Flexure Modulus (GPa)	48	19	57	23
ILSS (MPa)	77	43	63	32

Bibo et al. (1997) also examined the failure mode of non-crimp composites under tensile loading. In general the non-crimp and prepreg tape laminates failed in very similar ways, with multiple cracking in off-axis plies and delaminations between plies being recorded. However they observed that the warp knit yarn structure that joins the layers of in-plane yarns together, appeared effective in constraining the extent of delamination and longitudinal splitting in comparison to that observed in the unidirectional prepreg laminates. This improved resistance to interply failure and separation due to the through-thickness knitting yarns is also thought to be the cause for the improved interlaminar shear strength noted earlier.

7.3 INTERLAMINAR FRACTURE TOUGHNESS

It has been previously mentioned that the open nature of the knit architecture gives these fabrics the ability to nest very closely between individual fabric layers. Some knit architectures also consist of 2 or more fabric layers that are integrally connected by knitting yarns. Both these attributes of knitted fabrics will promote the formation of fibre bridging mechanisms that should enhance the fracture toughness of knitted composites.

Mode I fracture tests have been performed upon a range of E-glass/epoxy warp knitted composite materials (Huysmans et al., 1996), E-glass/epoxy weft knitted composites (Kim et al., 2000) and E-glass/vinyl ester weft knitted composites (Mouritz et al., 1999). In all cases the fracture toughness measurements of knitted composites were significantly higher than those of typical 2D woven, unidirectional or random mat composites. Huysmans *et al* (1996) measured Mode I fracture toughness levels of 5.5 to 6.5 kJ/m^2 for specimens of warp knitted E-glass/epoxy containing a tissue architecture. This is in direct comparison to typical values of 1.2 and 0.6 kJ/m^2 for woven and unidirectional E-glass/epoxy materials respectively. Mouritz et al. (1999) conducted an extensive comparison of the fracture toughness of milano weft knitted composites against a range of unidirectional, 2D woven, 2D braided, 3D woven and stitched E-glass/vinyl ester materials. The authors found that their toughness measurements for the knitted composites of up to 3.3 kJ/m^2 were not only approximately four times that of 2D woven materials but were also significantly higher than those measured for the 2D braided, stitched and 3D woven materials. Both Huysmans et al. (1996) and Mouritz et al. (1999) examined the fracture path of the knitted composite and found that the highly looped nature of the yarn architecture had forced the crack to follow a very tortuous path with extensive crack branching. They concluded that this combination of crack path deflection and crack branching is the likely cause of the high interlaminar fracture toughness.

Mouritz et al. (1999) also noted that the fracture toughness of the knit decreased when the stitch density of the knitted fabric increased. This was also observed by Kim et al. (2000) who measured the Mode I fracture toughness of milano weft knitted E-glass/epoxy composites at a range of tightness factors, this factor being directly proportional to the stitch density (see Table 7.6). They found that as the tightness factor increased the measured fracture toughness decreased in an approximately linear fashion. This effect is due to the fact that as the tightness factor (or stitch density) increases the knit architecture becomes progressively less open. When the fabric layers are placed together the fabric with a higher stitch density will nest, or intermingle, less than a fabric with an open structure. This lower degree of intermingling will result in a less tortuous crack path and thus a lower value of fracture toughness.

Table 7.6 Mode I fracture toughness of E-glass/epoxy weft knitted milano composites (from Kim et al., 2000)

Material	Fibre volume fraction (%)	Tightness factor	G_{1c} (kJ/m^2)
Milano 1	20.1	1.30	4.05
Milano 2	22.3	1.44	3.22
Milano 3	24.8	1.61	2.58
Milano 4	27.4	1.73	2.29

7.4 IMPACT PERFORMANCE

7.4.1 Knitted Composites

The superior properties of knitted composites in Mode I fracture toughness is also reflected in their overall impact performance. Leong et al (1998) examined the low- to medium-energy impact performance of an E-glass/epoxy, weft knitted milano material under drop-weight conditions. For the range of impact energies tested (up to 7.3 J/mm) they found that the damage area created within the knitted composite was essentially a circular region of very dense and complex microcracks. The diameter of this damage zone increased as you moved from the front face to the back face creating a trapezoidal shape. The authors found that the compression strength of the impacted composite was reduced by only 21% for high impact energies, implying that the knitted composite was very damage tolerant. Also, in comparison with composite specimens manufactured with uniweave reinforcements, the knitted composite was capable of absorbing a much higher proportion of the incident impact energy, 64% more than the uniweave composite at high impact energies.

This energy absorption capability has also been observed by Chou et al. (1992) who conducted notched Charpy impact tests upon E-glass/epoxy specimens of both weft knitted 1x1 rib and plain weave composite materials. They found that the absorbed impact energy of the plain weave composite was 68.3 kJ/m^2 whilst the knitted composite was at least 2.4 times better at 161.3 kJ/m^2. This ability of knitted composites to absorb substantially greater amounts of impact energy than woven materials would suggest them as ideal candidates for damage-prone structures or ones requiring a high energy absorption capability, such as crush members. This concept was investigated by Ramakrishna et al. (1993) when they examined the energy absorption capabilities of epoxy composite tubes reinforced with knitted carbon fabrics. The knit architectures used were weft knitted 1x1 rib structures with and without straight fibres laid in the course direction. The orientation of the inlay yarns along the tube axis allowed the specific energy absorption capability of the tube to reach 85 kJ/kg with only a total fibre volume fraction of 22.5%. This performance is encouraging when compared to the highest specific energy of 120 kJ/kg recorded for carbon/epoxy tubes with a fibre volume fraction of 45% (reported by Ramakrishna *et al* 1993).

The impact performance under drop weight conditions of knitted composites with regard to knit architecture has also been investigated by Khondker et al. (2000). They examined the impact resistance and tolerance of three different architectural styles of E-glass/vinyl ester weft knitted composites; milano, 1x1 rib and plain knit. For the three architectural styles, at similar stitch densities, the authors found that the damage area created at the same impact energy of 6 J/mm (an indication of the impact resistance) increased significantly from the plain knit (230 mm^2) through the milano (290 mm^2) to the 1x1 rib structure (350 mm^2). In a similar fashion the reduction in compression strength after impact, which gives an indication of the material's impact tolerance, also varied with knit architecture. Again the plain knit demonstrated the best damage tolerance, losing only 22% of its initial undamaged strength at an impact energy of 6 J/mm whilst the milano and 1x1 rib structures lost 27% and 32% respectively. It is not clear what aspect of the knit architecture gives the plain knit a superior impact performance over the milano and 1x1 rib structures.

Within each of the knit architectures the authors also examined the effect of varying the stitch density upon the composite impact performance. No significant effect on the impact resistance was seen within any of the three architectures even though the stitch density changes by a factor of two for each material. This lack of any conclusive change with stitch density was also observed for the impact tolerance of the knitted composites. This result is possibly expected as the undamaged compressive properties of knitted composites showed very little effect from variations in the loop parameters within any of the knit architectures examined. This was attributed to the dominant influence the matrix plays in the compressive properties, therefore given a similar extent of damage created within the composite, the remaining compressive strength should also be similar.

The behaviour of knitted composites under impact conditions is clearly a complex situation but what is worth emphasising is the ability of knitted composites to absorb large amounts of impact energy relative to other reinforcement forms and to suffer less relative degradation to their compressive performance. This is illustrated by Figure 7.9 (from Khondker et al., 2000) which compares the relative degradation in compression strength of typical composites manufactured with knitted, 2D braided, uniweave fabric and unidirectional tape reinforcements.

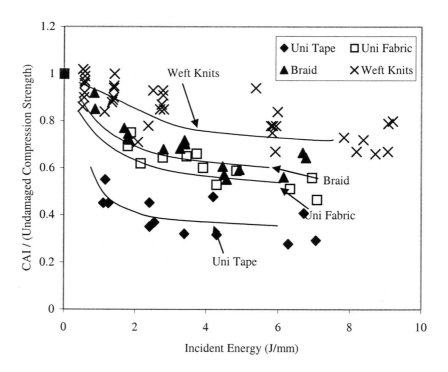

Figure 7.9 Compression-after-Impact (CAI) strength (normalised by the undamaged compression strength) of composite materials reinforced with various textile forms (from Khondker et al., 2000)

7.4.2 Non-Crimp Composites

The impact performance of carbon non-crimp composites was investigated by Bibo *et al* 1998 and compared with a unidirectional prepreg tape composite with the same resin system and lay up. No significant difference in the impact resistance was observed between the two material forms with similar damage areas being measured at the same impact energies, however examination of the damage patterns did reveal a slight variation. The damage sustained by the non-crimp composite appeared to be affected by the local presence of the knitting yarns and any undulations in the layered fabric, giving the damage a more complex appearance than the traditional delaminations and shear/transverse cracks observed in unidirectional prepreg tape composites. It is possible that these variations in the local fabric topography are acting as barriers to easy crack growth, forcing the crack to follow a more convoluted path, although this effect is not seen globally in the total measured damage area.

The residual compression strength after impact did not show any conclusive difference between the absolute values measured for the non-crimp and prepreg specimens. However, given that the undamaged compression strength of non-crimp composites was generally observed to be significantly lower than that of unidirectional prepreg tape materials, the authors claimed that the non-crimp composites exhibited a greater damage tolerance than the prepreg materials.

7.5 MODELLING OF KNITTED COMPOSITES

Given the complex nature of the knit architecture, accurately modelling the strength and stiffness performance of these materials is a very challenging task. Not withstanding this, a number of researchers have been developing modelling approaches to varying degrees of success and comparative studies of these approaches are contained in worthwhile reviews by Leong et al. (2000) and Huang et al (2000).

Historically there have been two general approaches to modelling the performance of knitted composites; Numerical (using FEM techniques) and Micromechanical. Although FEM is a very powerful tool for structural analysis the 3D complexity of the knit architecture and the sensitivity of FEM to boundary conditions make this approach both time-consuming and the applicability of the results potentially suspect. Micromechanical approaches have therefore become the more practical means of modelling the knitted composite.

The mechanical properties of a knitted composite will depend upon three things; the properties of the constituent materials, the overall fibre volume fraction, and the knit loop architecture. Of these three areas the most critical in the model development is the determination of the knit geometry. All of the models that have been developed for knitted composites start first by describing the Representative Volume Element (RVE) or unit cell of the knit architecture, ideally by some analytical function as discussed in Chapter 4. However, currently only the plain knit architecture can be specified by such a function (Leaf and Glaskin), other knit architectures must have their RVE's described through often time-consuming and difficult experimental measurements.

Once the RVE has been described the most simplistic approach reported has been to use the Krenchel model, which uses a combination of the rule of mixtures and a reinforcement efficiency factor to describe the elastic modulus. Predictions using this technique were generally lower than experimental values and its one-dimensional approach limits its capability to predict the full set of elastic constants.

More complex methods generally involve partitioning the RVE into a number of infinitesimal elements (sub-elements), the properties of which are analysed by means of a unidirectional micromechanics model in the local coordinate systems. A tensor transformation is then used to transform the sub-elements from local coordinates to a global one and an averaging scheme, normally either iso-strain (Voight method), iso-stress (Reuss method) or a variation of these, is used to obtain the overall stiffness matrix of the RVE.

There are many micromechanical models in the literature that can be used to define the unidirectional properties of the subelements. Two of these that have been used in the modelling of knitted composites are the Chamis model, which can only be used to model the elastic properties, and the Bridging Matrix model, which has the capability to model the stress-strain behaviour of the composite up to failure.

A comparison of a number of these modelling approaches was made by Huang *et al* (2000) for the prediction of the tensile properties of an E-glass/epoxy composite reinforced by a single layer of plain weft knitted fabric. The results of that comparison showed that there is no one combination of micromechanical model and averaging scheme that currently gives reasonable predictions for the elastic properties and failure strengths. In general the errors in the predictions ranged between 15% to 29% from the measured values, and often a modelling scheme whose prediction was close for one particular property produced a very inaccurate prediction for another property.

More recent modelling work (Huang et al., 2001; Huysmans et al., 2001) is showing promise for the accurate prediction of the mechanical performance of knitted composites but a substantial amount of progress is needed before a robust, accurate modelling approach is available.

7.6 SUMMARY

Knitted fabrics hold a great deal of potential for the manufacture of specific types of composite components. No other textile reinforcement is as capable as knitted fabric is, of being formed or directly manufactured into complex shapes. Their excellent impact performance would appear to make them ideal for service conditions where energy absorption or damage tolerance was critical. A special sub-group of knitted fabrics, known as Non-crimp Fabrics, is also capable of manufacturing parts with very high in-plane mechanical performance at a reduced manufacturing cost and is a prime material candidate to replace conventional prepreg materials in future aircraft.

As with many of the 3D textile reinforcements described here, the mechanical performance of knitted fabrics is a very complex and not well understood issue. Excepting non-crimp materials, knitted composites have in-plane mechanical properties that lie between that of random mats and traditional 2D weaves, but these properties can be dramatically changed by the knit architecture and the degree of stretch within the knit. The generation of a database of knitted composite properties and the development of models to understand and predict these properties are still in their infancy relative to the other forms of 3D reinforcement. Further progress in these two areas is required before knitted fabrics will become a more commonly used reinforcement in composite structures.

Chapter 8

Stitched Composites

8.1 INTRODUCTION TO STITCHED COMPOSITES

Stitching has been used with notable success in the manufacture of advanced 3D composite materials since the early 1980s. The stitching of composites was first investigated by the aircraft industry to determine whether it could provide through-thickness reinforcement to FRP joints. The aircraft industry investigated the feasibility of stitching wing-to-spar and single-lap composite joints to increase the failure strength and reduce the likelihood of sudden catastrophic failure (Cacho-Negrete, 1982; Holt, 1992; Lee and Liu, 1990; Sawyer, 1985; Tada and Ishikawa, 1989; Tong et al., 1998; Tong and Jain, 1995; Whiteside et al., 1985). The investigations revealed that the strength of stitched joints was superior to composite joints made using conventional joining techniques such as adhesive bonding and co-curing. The failure strength of stitched joints was found to match or in some cases, exceed the strength of composite joints reinforced with metal rivets. Despite the great potential benefits offered by stitching, the aircraft manufacturing industry has been slow to adopt stitching as a method for reinforcing composite joints. However, stitched joints may become common in next-generation aircraft.

An important outcome of the early stitching work on composite joints is that it sparked great interest in the stitching of a wide variety of FRP materials. While the implementation of stitched joints into aircraft is proving to be a slow process, stitching is rapidly becoming a popular technique for reinforcing composite panels for use in aircraft. This growing popularity is due mainly to two attributes of the stitching process. Firstly, stitching is a cost-effective method for joining stacked fabric plies along their edges to make the preform easier to handle prior to liquid moulding. Without stitching or some other type of binding, stacked plies often slip during handling that can cause fibre distortions and resin-rich regions in the composite. The second benefit of stitching is that it can improve the delamination resistance and impact damage tolerance of composites.

The benefits gained by stitching are spurring the development of a wide variety of stitched composite components. As described in Chapter 1, aircraft manufacturers are evaluating stitching for possible use in wing skin panels and fuselage sections (Bannister, 2001; Bauer, 2000; Beckworth and Hyland, 1998; Brown, 1997; Deaton et al., 1992; Dexter, 1992; Hinrichen, 2000; Jackson et al., 1992; Jegley and Waters, 1994; Kullerd and Dow, 1992; Markus, 1992; Mouritz et al., 1999; Palmer et al., 1991; Smith et al., 1994; Suarez and Daston, 1992). It is expected that the damage tolerance of composite structures will be improved dramatically with stitching, thereby increasing the structural reliability of aircraft. For example, stitching is being considered for stiffening the centre fuselage skin of the Eurofighter (Bauer, 2000) and the rear pressure bulkhead of the Airbus A380 aircraft (Hinrichsen, 2000). Stitching is also being assessed for use in automobile components prone to impact, such as bumper bars, floor panels and door members (Hamilton and Schinske, 1990). The feasibility of using stitching for other applications, such as in boats, civil structures and medical prostheses, has not yet been explored in detail (Mouritz et al., 1999). As the technology is developed further stitched composites are likely to be used in a wide range of applications.

The fabrication, mechanical properties, delamination, impact damage performance and joining performance of stitched composites are described in this chapter. The stitching textile technologies that are used to fabricate stitched composites are outlined in the next section. Included in the section is a description of the different 3D fibre architectures that can be produced with stitching. Following this, the effect of stitching on the in-plane mechanical properties and failure mechanisms of composites are described in Section 8.3. This includes a description of the tension, compression, bending, creep and fatigue properties of stitched composites. The interlaminar properties and delamination resistance of stitched properties are then described in Section 8.4. This includes an examination of the modes I and II interlaminar fracture properties and delamination toughening mechanisms of stitched composites, and a description of analytical models that have been developed to predict the delamination resistance of stitched materials. The effect of stitching on the impact damage tolerance of stitched composites is examined. Finally, the use of stitching for the reinforcement and stiffening of composite joints is outlined in Section 8.6.

8.2 THE STITCHING PROCESS

The stitching process basically involves sewing high tensile thread through stacked ply layers to produce a preform with a 3D fibre structure. A schematic of the 3D fibre structure of a stitched composite is illustrated in Figure 2.31. It is possible to stitch a thin stack of plies using conventional (household) sewing machines. Although it is more common to stitch using an industrial-grade sewing machine that has long needles capable of piercing thick preforms. The largest sewing machines for stitching composites have been custom built for producing large panels up to 15 m long, nearly 3 m wide and 40 mm thick. Figure 8.1 shows the largest sewing machine yet built, and this is used for stitching the preforms to aircraft wings panels (Beckwith and Hyland, 1998; Brown, 1997; Smith et al., 1994). Many of the latest machines have multi-needle sewing heads that are robotically controlled so that the stitching process is semi-automated to increase sewing speeds and productivity.

Stitched composites are similar to 3D woven, braided and knitted composites in that the fibre structure consists of yarns orientated in the in-plane (x,y) and through-thickness (z) directions. A feature common to 3D woven, braided and knitted materials is that the in-plane and through-thickness yarns are interlaced at the same time during manufacture into an integrated 3D fibre preform. The stitching process, on the other hand, is unique in that the stitched preform is not an integral fibre structure. The through-thickness stitches are inserted into a traditional 2D preform as a secondary process following lay-up of the plies.

Stitching can be preformed on both dry fabric and uncured prepreg tape. Stitching most types of fabric is relatively easy because the needle tip can push aside the dry fibres as it pierces the preform. Sewing prepreg tape can be more difficult because the inherent tackiness of the uncured resin matrix impedes the needle action. The materials most often used as the reinforcing threads for stitching are carbon, glass and Kevlar yarns, although it is possible to sew with other types of fibrous materials including Spectra® and high strength thermoplastics. The yarns can be sewn into the preform in a variety of patterns, with the most common types being the lock stitch, modified lock stitch and chain stitch. These three stitch types are shown in Figure 8.2 (Morales, 1990). The standard lock and chain stitches are used occasionally, but the most popular stitch style is the modified lock stitch in which the loops crossing the needle and bobbin threads are formed at one surface of the composite to minimise in-plane fibre distortions inside the material.

Figure 8.1 Large stitching machine used to stitch composite wing panels (From Beckwoth & Hyland, 1998).

When composites are stitched the through-thickness threads can be inserted in any number of patterns. Examples of stitch patterns used to reinforce composites are shown in Figure 8.3, and of these the most popular pattern is horizontal stitching (Dransfield et al., 1994). During the stitching process the threads are usually placed close together to ensure high damage tolerance, and most composites are reinforced with 1 to 25 stitches per cm². This is equivalent to a fibre volume content of stitched threads of about 1% to 5%. This is a similar volume content for the through-thickness reinforcement in many 3D woven, braided and knitted composites. It is often difficult to stitch composites at higher densities because of the excessive amount of damage caused to the preform.

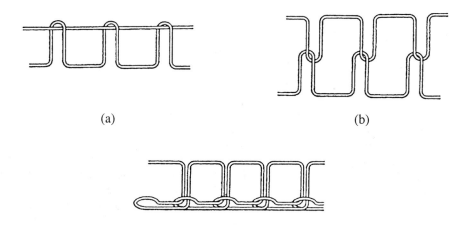

(a)   (b)

(c)

Figure 8.2 Illustrations showing the configuration of the (a) modified lock stitch, (b) lock stitch and (c) chain stitch (From Morales, 1990).

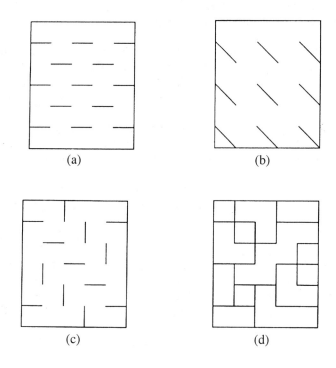

(a)   (b)

(c)   (d)

Figure 8.3 Illustrations showing (a) straight, (b) diagonal, (c) zig-zag and (d) cross stitching (Reproduced from Dransfield et al. , 1994).

The damage to preforms caused by stitching is one of the major drawbacks of this technique. Stitching can cause many different types of damage, and these are shown in Figure 8.4. The different damage types are:

Fibre breakage: Breakage occurs from abrasion generated by the needle and stitch yarn sliding against the fibres during the stitching process. Breakage in prepreg tape can also occur by the needle tip crushing the fibres, which cannot be easily pushed aside by the needle because of the resin matrix. Fibre breakage in a dry fabric preform in shown in Figure 8.4a.

Fibre misalignment: Significant misalignment of the in-plane fibres occurs because they are distorted around the needle and stitch thread. The amount of distortion to fibres depends on the fibre density, stitch yarn thickness and, in some cases, stitching density, and maximum misalignment angles of between 5° and 20° have been measured (Mouritz and Cox, 2000; Mouritz et al., 1996; Reeder, 1995). A schematic diagram and photograph of fibre misalignment is shown in Figures 8.4b and 8.4c.

Fibre crimping: Crimping occurs by the stitches drawing the fibres at the surface into the preform. The effect of fibre crimping is illustrated in Figure 8.4d, and the severity of crimping increases with the line-tension on the stitching yarn.

Resin-rich regions: The crimping and misalignment of fibres in preforms leads to the formation of small regions with a low fibre content around the stitches(Figure 8.4e). This leads to the formation of resin-rich regions when the preform is consolidated into a composite using liquid moulding processes.

Stitch distortions: The stitches can be distorted by heavy compaction of the preform using liquid moulding, hot pressing or autoclaving techniques (Rossi, 1989). This type of damage is shown in Figure 8.4f.

Microcracking: Cracking of the resin near the stitches can occur due to thermally-induced strains arising from the mismatch in the coefficients of thermal expansion of the stitches and surrounding composite material (Farrow et al., 1996; Hyer et al., 1994). In some stitched materials, this can cause debonding between the stitches and composite.

Compaction: Applying a high tensile load to the thread to ensure it is taut during stitching can compact the preform plies. As a result, consolidated stitched composites can have fibre volume fractions that are several percent higher than expected.

Not all stitched composites contain all the different types of damage listed above. The most common types of damage to stitched composites are fibre breakage, misalignment and crimping.

In addition to damage to the composite, the stitch thread itself can be damaged. Damage to fibres in the threads occurs by twisting, bending, sliding and looping actions as the thread passes through the sewing machine and formed into a stitch. The damage can be significant and cause a large loss in strength (Dransfield, 1995; Jain and Mai, 1997; Morales, 1990). For example, Morales (1990) found that the tensile strength of Kevlar thread fell from 4790 MPa to 3706 MPa after stitching. The situation can be even worse when stitching with carbon thread, when a reduction in strength from 3500 MPa to only about 1550 MPa can occur (Morales, 1990).

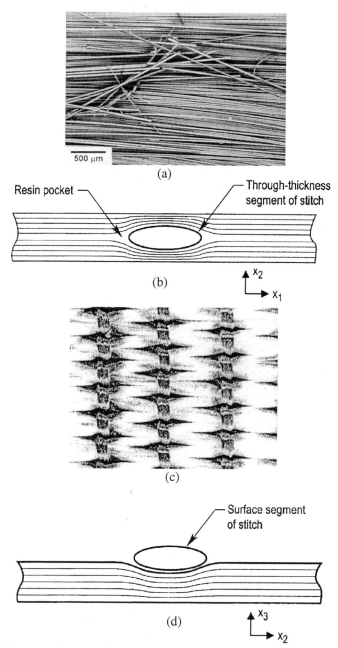

(a)

Resin pocket ─   ─ Through-thickness
                                               segment of stitch

(b) x_2
                             x_1

(c)

                             ─ Surface segment
                               of stitch

(d) x_3
                             x_2

Figure 8.4 Photographs and illustrations of stitching damage. (a) Dry woven fabric showing broken fibres caused by stitching. (b) Schematic and (c) photograph showing the local misalignment of fibres around a stitch. From Mouritz and Cox (2000) and Wu and Liau (1994), respectively. (d) Schematic of fibre crimping caused by stitching. (From Mouritz and Cox, 2000).

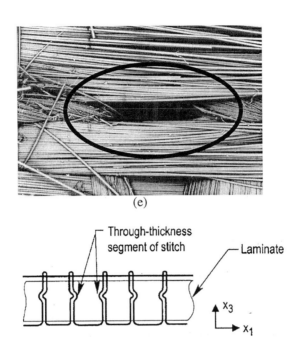

(e)

(f)

Figure 8.4 (Continued) Photographs and illustrations of stitching damage. (e) A region of low fibre content due to stitching is shown within the circle, and this develops into a resin-rich region when the composite is consolidated. (f) Schematic of the distortion to stitches caused by heavy compaction of the preform (From Mouritz and Cox, 2000).

8.3 MECHANICAL PROPERTIES OF STITCHED COMPOSITES

8.3.1 Introduction

The application of stitched composites to load-bearing structures on aircraft, such as wing skin panels and fuselage sections, requires an in-depth understanding of their mechanical properties and failure mechanisms. The mechanical property data is needed to validate design codes for stitched composites to be used in high performance structures. In this section the effect of stitching on the tensile, compressive, flexure, interlaminar shear, creep and fatigue properties of composite materials will be described. It will be shown that there is not a complete understanding of the effect of stitching on the mechanical properties of composites. In addition, models for predicting changes to the properties of composites due to stitching are not fully developed.

Until a strong modeling capability combined with a comprehensive database of mechanical properties for stitched composites is achieved, then the certification and application of these materials to primary aircraft structures will be difficult. Despite some shortcomings in our knowledge, there is much about the mechanical properties of stitched composites that is understood.

8.3.2 Tension, Compression and Flexure Properties of Stitched Composites

The tension, compression and bending modulus and strength are material properties of great engineering importance in load-bearing structures, and therefore these properties have been measured for many types of stitched composites, including carbon-, glass- and Kevlar fibre laminates. Large databases for the tension, compression and flexural properties are now available for most of the main engineering composites, including carbon/epoxy laminates used in aircraft. However, reliable models for predicting the in-plane mechanical properties of stitched composites are not available.

A review of the published mechanical property data for stitched composites shows apparent contradictions between materials (Mouritz and Cox, 2000; Mouritz et al., 1999). The data indicates that most stitched composites have slightly lower tension, compression and flexural properties than their equivalent unstitched laminate, although some stitched materials exhibit no change or a modest improvement to their mechanical properties. For a few materials the properties are dramatically improved or severely degraded by stitching. The apparent contradictions are shown in Figure 8.5, which compares the tensile modulus and strength for two composites stitched under identical conditions (Kan and Lee, 1994). The Young's modulus for the glass/polyester decreases with increasing stitch density whereas the modulus for the Kevlar/PVB-phenol increases erratically with stitch density. The tensile strength for the glass/polyester also drops rapidly with increasing stitch density while the strength for the Kevlar/PVB-phenol increases slightly with stitch density before decreasing. Similar contradictions occur for the compression and flexure properties of stitched composites.

Mouritz and Cox (2000) analysed tension, compression and flexural property data from the literature for a variety of carbon-, glass-, Kevlar- and Spectra-fibre reinforced polymer composites stitched over a range of area densities (from 0.2 to 25 stitches/cm^2). The composites were stitched with different thread materials using lock, modified lock and chain stitches. The mechanical property data collected by Mouritz and Cox (2000) is plotted in Figures 8.6 to 8.8. The data is plotted as normalised Young's modulus (E/E_o) and normalised strength (σ/σ_o) against stitch density for tension, compression and flexure. The subscripts *t*, *c* and *f* to (E/E_o) and (σ/σ_o) represent tension, compression and flexure, respectively. The normalised Young's modulus is the modulus of the stitched composite (E) normalised to the modulus of the equivalent unstitched material (E_o) subject to the same load condition. Similarly, the normalised strength is the strength of the stitched composite (σ) divided by the strength of the unstitched laminate (σ_o) for the same load condition. In the figures, CFRP represents carbon fibre reinforced polymer, GFRP is glass reinforced polymer, KFRP is Kevlar reinforced polymer, and SFRP is Spectra reinforced polymer laminate.

With the exception of a few outlying data points, it is shown in Figures 8.6 to 8.8 that stitching improves or degrades the modulus and strength by no more than ~20%. Within this variance, there is no clear correlation between the change to the mechanical properties and stitch density. This implies that tension, compression and flexural failure is not determined by the collective action of many stitches but rather that a single stitch or a small number of stitches and the damage arising from them (eg. distortion and breakage of fibres) can determine strength.

This data trend is of practical significance because it shows that the tension, compression and flexure properties for most composites will be changed by less than 20% regardless of the amount of stitching. However, the impact damage resistance and post-impact mechanical properties can be improved with large amounts of stitching (see Section 8.5). Therefore, it appears that composites can be heavily stitched to provide maximum impact damage tolerance without reducing the in-plane mechanical properties any more than would occur with low density stitching.

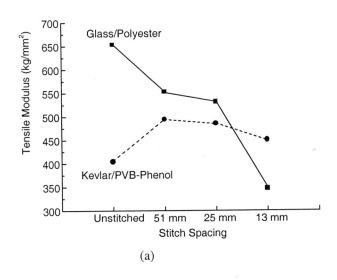

(a)

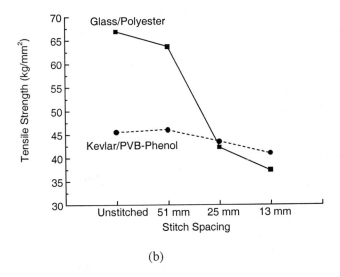

(b)

Figure 8.5 The effect of stitch density on the (a) Young's modulus and (b) tensile strength of a glass/polyester and Kevlar/PVB-phenol composite that were stitched under identical conditions. Data from Kang and Lee (1994).

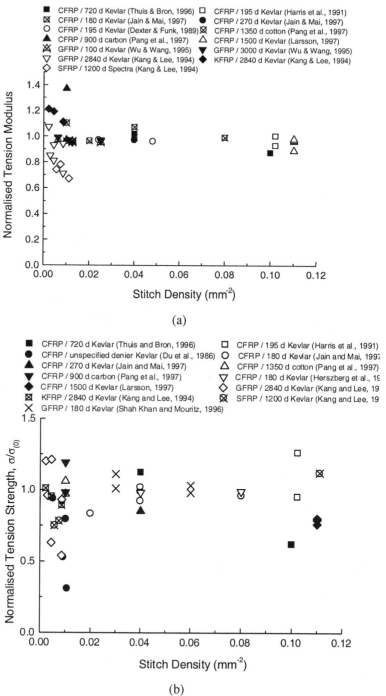

Figure 8.6 Plots of (a) normalised Young's modulus $(E/E_o)_t$ and (b) normalised tensile strength $(\sigma/\sigma_o)_t$ against stitch density (Mouritz and Cox, 2000).

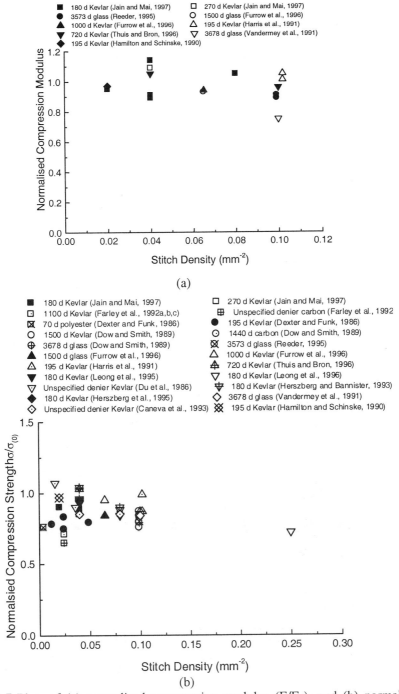

Figure 8.7 Plots of (a) normalised compression modulus $(E/E_o)_c$ and (b) normalised compression strength $(\sigma/\sigma_o)_c$ against stitch density (Mouritz and Cox, 2000).

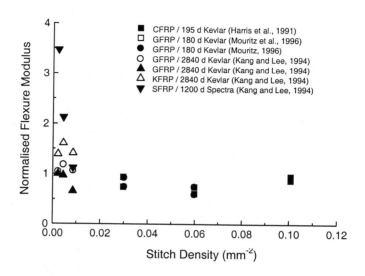

(a)

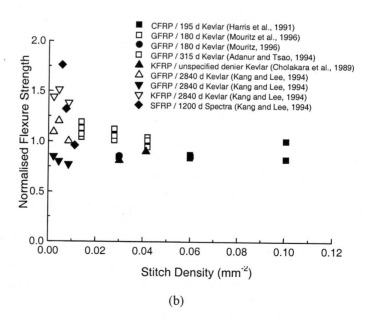

(b)

Figure 8.8 Plots of (a) normalised flexure modulus $(E/E_o)_f$ and (b) normalised flexure strength $(\sigma/\sigma_o)_f$ against stitch density (Mouritz and Cox, 2000).

Simple rule-of-mixtures theory can be used to explain the small improvements (<10%) to the Young's modulus and strength of stitched composites. It is well known that the modulus and strength of composites are dependent on the volume fraction of in-plane fibres aligned in the load direction. Stitching with a high tensile force on the yarn so the thread is taut can raise the fibre volume content by several percent by compacting the preform. The fibre volume fraction can be increased further by compaction of the stitched preform during consolidation inside a closed mould. It has been suggested by Mouritz and Cox (2000) that any modest increase to the Young's modulus and strength of stitched composites is probably due to a small increase in the fibre volume content due to compaction of the in-plane fibres by stitching. Unfortunately, however, most researchers who find an improvement to the mechanical properties of composites after stitching do not report the fibre volume contents for the stitched and unstitched materials. Therefore, while it is likely that compaction is the cause for the small improvements to modulus and strength, this has not been confirmed by experiment.

A few stitched composites display remarkably high mechanical properties, particularly under flexural loading. It is seen in Figure 8.8 that the flexural modulus and strength values for some stitched composites are up to 3.5 and 1.75 times higher than the unstitched laminate, respectively. Such large improvements to the flexural properties cannot be due solely to fibre compaction from stitching, which usually increases the fibre volume content by a few percent. The mechanism responsible for the large increase to the flexural properties of stitched composites has not been determined. However, a study by Chang et al. (1989) found that some unstitched composites fail in bending by delamination cracking, but when these materials are stitched the delamination cracks are suppressed and this increases the flexural properties.

The reduction to the tensile modulus and strength of stitched composites is attributed mostly to three types of damage caused by stitching. These are fibre breakage (see Figure 8.4a), distortion and crimping of the in-plane fibres causing misalignment from the load direction (see Figures 8.4b-8.4d). These types of damage are usually localised to a small region surrounding the stitches, and therefore their effect on the modulus and strength is expected to be modest. Because the number and size of the regions containing misaligned fibres is small, preliminary modelling by Mouritz and Cox (2000) indicate that the reduction to the Young's modulus of stitched composites is modest (usually less than about 5%). This prediction shows reasonable agreement with the observed reduction of up to 20% in many stitched composites.

The mechanism of compression failure of stitched composites is a complex process, and is believed to be a competition between different failure modes. Many types of unstitched composites, and in particular prepreg laminates, fail under uniaxial compression loading by delamination cracking between the plies. Delamination is the most common compression failure mechanism in unstitched materials, and it is often initiated by edge stresses or a pre-existing defect (such as a void or crack). Delamination failure can be suppressed in stitched composites by the stitches stopping Euler buckling of the delaminated plies. Farley et al. (1992a; 1992b; 1993c) observed that with delamination suppressed in stitched composites, the compressive failure mechanism changes to kinking of the most severely crimped load-bearing fibres. The damage mechanism leading to a kink failure is believed to start at the surface of stitched materials where the crimping of the fibres is the most severe (see Figure 8.4d). Under compression loading, axial shear stresses rapidly rise in a heavily crimped fibre bundle, which cause failure of the polymer matrix within the bundle via microcracking and crazing. This damage permits further rotation of the fibre bundle until it becomes axially unstable and a kink band is formed. Mouritz and Cox (2000) suggest that a kink band will initiate in a single region of a stitched composite that has suffered the highest degree of fibre distortion. Since the spread of a kink band is usually unstable, a stitched composite may fail before other kink bands form near stitches that have caused less distortion. This process accounts for the observation in Figure 8.8b that the reduction to compression strength is not affected by stitch density, because even the most lightly stitched materials still have severely crimped fibres.

Stitching is expected to have a beneficial effect on the flexural properties of composites because it suppresses the growth of delaminations formed under bending, and thereby increases the strength. However, this is believed to be offset by damage incurred with stitching, particularly the distortion and breakage of fibres, which lowers the flexural properties. In some types of stitched composites, the distortion of fibres close to the constrained laminate surface can cause bending-induced compression failure at a reduced flexural stress. In other stitched laminates, the clusters of broken fibres close to the stitches leads to fibre fracture on the tensile side of a flexural specimen. From existing data and limited observations, it appears that there is competition between the failure mechanisms within stitched composites. That is, stitching suppresses delamination cracking which can raise the strength, but stitching also facilitates compression failure and tensile rupture that can lower the strength. The competition between these different mechanisms is probably a close one, and this would account for the modest reduction to the flexural properties with stitching.

8.3.3 Interlaminar Shear Properties of Stitched Composites

The interlaminar shear properties of stitched composites have not been extensively evaluated, and as yet the effect of stitching on these properties is not fully understood. From published research it appears that the interlaminar shear strength, like the tension, compression and flexure properties, can be improved and degraded by stitching (Mouritz et al., 1997). For example, Figure 8.9 shows the effect of stitch density on the interlaminar shear strengths for Kevlar/epoxy and carbon/epoxy composites that have been stitched using Kevlar yarn (Jain and Mai, 1997; Kang & Lee, 1994).   The interlaminar shear strength of the Kevlar/epoxy composite increases steadily with stitch density, and this is attributed to the suppression of interlaminar cracking by the stitches. The strength of the carbon/epoxy composite, on the other hand, drops slightly when stitched, although the strength does not appear to be affected significantly by the stitch density.

Figure 8.10 presents interlaminar shear strength data for a variety of composites plotted against stitch density. The data was collected from various published sources by Mouritz and Cox (2000). The normalised interlaminar shear strength (τ/τ_o) is defined as the strength of the stitched composite (τ) divided by the strength of the equivalent unstitched laminate (τ_o). The range of enhancement or degradation of interlaminar shear strength is about 15-20%, which is similar to the improvement or reduction to the tension, compression and flexure properties of stitched composites.

The improvement to the interlaminar shear strength is probably due to the stitches inhibiting the delamination crack growth process (Cholarkara, 1989; Mouritz and Cox, 2000). It is well known that as an interlaminar crack grows through a laminate, a zone forms behind the crack tip where stitches bridge the delamination. This is known as a stitch bridging zone, and the stitches apply bridging tractions to the delamination and thereby impede crack growth. The bridging zone can extend a long way behind the crack tip, usually several tens of millimeters or more, and is highly effective in slowing the rate of crack growth and thereby increasing the interlaminar shear strength. The bridging mechanism for increasing the interlaminar strength of composites is described in more detail in Section 8.4.

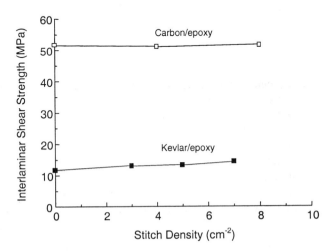

Figure 8.9 Effect of stitch density on the apparent interlaminar shear strength, τ, of Kevlar/epoxy and carbon/epoxy composites (data for the Kevlar/epoxy and carbon/epoxy is from Cholarkara et al. (1989) and Jain and Mai (1997) respectively).

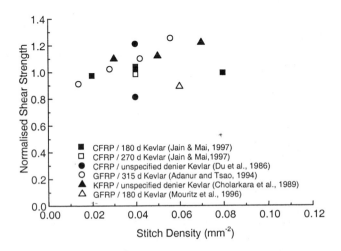

Figure 8.10 Plot of normalised interlaminar shear strength (τ/τ_o) against stitch density. The denier and material used for the stitching are as given (Mouritz and Cox, 2000).

The mechanism responsible for the slight reduction to the interlaminar shear strength of stitched composites is not fully understood. Some studies have observed that pure interlaminar shear cracking along the mid-plane of stitched composites does not always occur. Instead, failure occurs by cracking of the most highly distorted fibres close to the stitches rather than by pure interlaminar shear cracking. It is proposed that this change to the failure mechanism from pure shear cracking to fibre cracking is the likely cause for the lower interlaminar shear strength of some stitched composites (Mouritz et al., 1996).

8.3.4 Creep Properties of Stitched Composites

The creep properties of stitched composites have not been examined in detail despite the importance of creep strength in materials operating at elevated temperature. Pang and colleagues (Bathgate et al., 1997; Pang et al., 1997; Pang et al., 1998) investigated the effect of stitching on the creep performance of carbon/glass hybrid composites. Figure 8.11 shows the effect of creep strain for the hybrid composite in the unstitched condition and when stitched with carbon or cotton yarn. It is shown that the amount of creep is reduced dramatically with stitching, and the carbon stitching provides the highest creep resistance. It was found by Pang et al. that at a given time, the applied tensile stress required to induce the same amount of creep in stitched composites can be at least twice that for the equivalent unstitched laminate. It is proposed that stitching improved the creep resistance by increasing the interlaminar strength. Further research into creep rates and mechanisms of other types of engineering composites is needed before a complete understanding is gained of the creep properties of stitched materials.

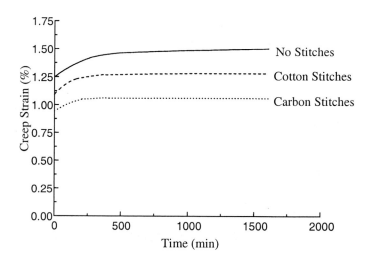

Figure 8.11 Creep curves for a carbon-glass hybrid composite in an unstitched condition and stitched with cotton and carbon threads (Adapted from Bathgate et al., 1997).

8.3.5 Fatigue Properties of Stitched Composites

An understanding of the fatigue endurance of stitched composites is essential because of their potential applications in structures that are subject to long-term fluctuating loads, such as aircraft wing panels and beams for civil infrastructure. The fatigue performance of stitched composites has been determined for a variety of cyclic load conditions, including repeated compression-compression and tension-tension loading.

The effect of stitching on the fatigue properties of composites under cyclic compression loads has been examined in great detail, particularly the stitched quasi-isotropic carbon/epoxy laminates that may be used in future aircraft (Dow and Smith, 1989; Furrow et al., 1996; Lubowinski and Poe; 1987; Portanova et al., 1992; Vandermey et al., 1991). Stitching is generally found to reduce the compression fatigue life of composites, with no study yet to observe an improvement to fatigue performance. A typical example of the effect of stitching on the compression fatigue-life (or S-N) curve for a carbon/epoxy composite is shown in Figure 8.12 (Portanova et al., 1992). It is seen that stitching reduces the fatigue life of the composite, and the reduction is due mainly to fibre distortion and crimping caused by stitching. Under cyclic compression loading it is observed that fatigue damage initiates at the fibres close to those stitches that have experienced the greatest out-of-plane distortion. It is believed that the fibre distortion promotes the early formation of fatigue-induced kink bands, which gradually rotate to even greater misalignment angles under continued fatigue loading until catastrophic failure.

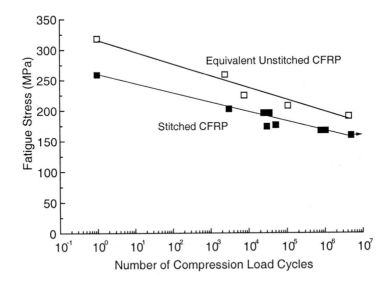

Figure 8.12 S-N curves for stitched and unstitched carbon/epoxy composites fatigued under compression-compression loading (Data from Portanova et al., 1992).

A simple model for estimating the compression fatigue life of stitched composite has been proposed by Mouritz and Cox (2000). They observed that the compression S-N curves for stitched and unstitched composites have slopes that are essentially indistinguishable beneath the experimental scatter, as seen for example in Figure 8.12. The only significant difference between the S-N curves for stitched and unstitched composites is the initial knock-down in static compression strength suffered by a stitched material. As shown in Section 8.3.3, the knock-down in compression strength due to stitching is usually under 20%. Based on this difference, Mouritz and Cox (2000) propose that the S-N curve for a stitched composite subject to compression fatigue can be estimated using Basquin's law:

$$S = \sigma_o - m \log_{10} N \qquad\qquad (8.1)$$

where S is the maximum applied compressive fatigue stress, N is the number of load cycles, σ_o is the static compressive strength of the stitched composite, and $-m$ is the slope of the S-N for the unstitched laminate.   Figure 8.13 shows fatigue life data for two stitched composites determined experimentally by Portanvona et al. (1992) and Vandermey et al. (1991). The solid S-N curves in Figure 8.14 show the theoretical fatigue life determined using equation (8.1). It is seen that the fatigue curve of stitched composites can be accurately predicted using the model. An appealing feature of the model is its simplicity; the compression S-N curve for a stitched composite can be determined from two simple tests: (1) a static compression test on the stitched composite to measure the compressive strength, σ_o, and (2) a compression-compression fatigue test on the unstitched composite to determine the slope of the S-N curve, $-m$.

Stitching can also degrade the tension-tension fatigue resistance of composites (Aymerich et al., 2001; Herszberg et al., 1997; Shah Khan and Mouritz, 1996, 1997). For example, Figure 8.14 compares fatigue-life curves for an unstitched and stitched composite subject to zero-to-tension fatigue loading (Shah Khan and Mouritz, 1997). An examination of the fatigue damage mechanisms of the composites shown in Figure 8.14 reveals that the stitched laminate started showing evidence of fatigue-induced damage close to the stitches at a relatively low number of load cycles. It is believed that the distortion and clusters of broken fibres caused by stitching act as sites for the early growth of fatigue-induced damage that ultimately leads to complete failure of the stitched composite. Aymerich et al. (2001) have found, however, that the tensile fatigue performance is only degraded in fibre dominated composites, such as with a $[0]_s$ or $[\pm45/0/90]_s$ stacking sequence. The fatigue performance of matrix dominated composites (eg. $[\pm30/90]_s$) is improved by stitching because the threads are effective in arresting or delaying the delamination crack growth under tensile fatigue loading.

Despite the knowledge of the fatigue performance of stitched composites in compression-compression and tension-tension loading, further research into fatigue is needed. The conditions under which stitching is beneficial or detrimental to the tensile fatigue endurance of composites still needs to be resolved. The effects of the various fatigue conditions (eg. R-ratio, load frequency) and stitching conditions (eg. yarn thickness, stitch density) on the fatigue endurance and fatigue damage mechanisms of the main engineering composites, particularly carbon/epoxy, is required. Research into the fatigue performance of stitched composites subject to reversed (compression-tension) cyclic loading is also needed.

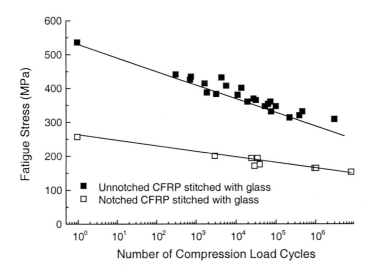

Figure 8.13 A comparison of measured fatigue data points against theoretical S-N curves for two types of stitched carbon/epoxy composites fatigued under compression-compression loading (Data for the upper and lower data points are from Vandermey et al. (1991) and Portanova et al. (1992), respectively).

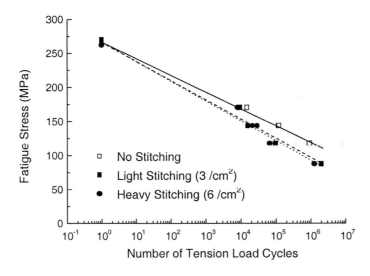

Figure 8.14 S-N curves for an unstitched and two stitched glass/vinyl ester composites fatigued under zero-to-tension loading (Data from Shah Khan and Mouritz, 1996).

8.4 INTERLAMINAR PROPERTIES OF STITCHED COMPOSITES

8.4.1 Mode I Interlaminar Fracture Toughness Properties

A key benefit of stitching is a vast improvement to the delamination resistance of composites. Stitching is a remarkably effective technique for increasing the interlaminar fracture toughness of laminates under mode I loading conditions (ie. crack opening). The toughening effect of stitching is shown in Figure 8.15, which shows mode I crack growth resistance (or R-) curves for a stitched and unstitched composite. This figure shows the R-curve for a glass/vinyl ester composite that has been stitched with Kevlar yarn. The R-curve behaviour of other types of stitched composites is similar to that shown in Figure 8.15. The mode I strain energy needed to start the growth of a delamination in a stitched composite is normally the same as the unstitched laminate. However, the R-curve for the stitched composite rises rapidly with crack length up to about 20 mm due to the increased toughening provided by the stitches. At longer crack lengths, the curve becomes relatively constant, and this is taken to be the steady-state interlaminar fracture toughness of the stitched composite (G_{IR}). At this stage the delamination resistance of the stitched composite is much higher than the unstitched laminate.

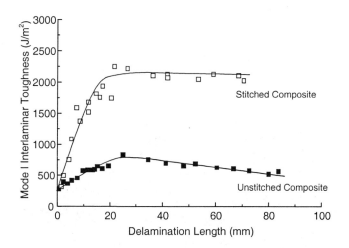

Figure 8.15 Mode I R-curves for a stitched and unstitched glass/vinyl ester composite.

The improved delamination resistance is due to the bridging action of the stitches, which exert a closure traction force that lowers the tensile strain acting on the crack tip. A schematic illustration is presented in Figure 8.16 of the interlaminar toughening provided by stitching under mode I loading. Ahead of an advancing delamination the stitches are not noticeably damaged or deformed, and when a crack reaches the stitches it passes around without causing the threads to break. Thereby, a zone is developed behind the crack tip where unbroken stitches bridge across the delamination. Within this so-called 'stitch bridging zone' the stitches are able to support a large amount of the applied stress by elastic stretching because of the high modulus and strength of the thread material. As a result, the tensile stress acting on the crack tip is reduced, and this slows the crack growth rate. A photograph of a stitch bridging a delamination crack is shown in Figure 8.17.

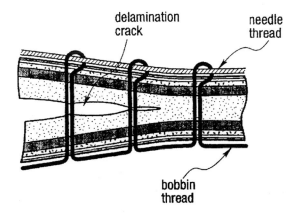

Figure 8.16 Schematic diagram of the mode I interlaminar toughening process in stitched composites (From He and Cox, 1998).

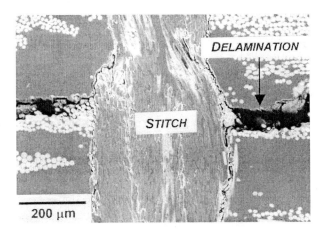

Figure 8.17 Scanning electron micrograph showing a stitch bridging a delamination crack (from Watt et al., 1997).

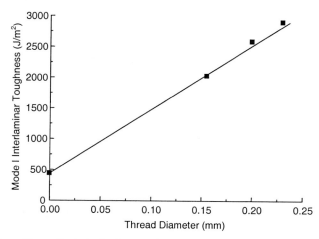

Figure 8.20 The effect of stitch denier on the mode I interlaminar fracture toughness of a carbon/epoxy composite. The composite was stitched with Kevlar thread to an areal density of 4 /cm^2 (Data from Jain & Mai, 1997).

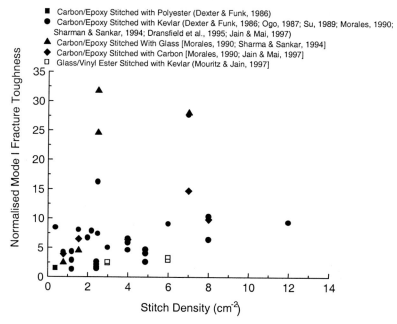

Figure 8.21 The effect of stitch density on the normalised mode I interlaminar fracture toughness (G_{IR}/G_{Ic}) for a variety of stitched composites (Mouritz and Jain, 1999).

In Figure 8.21, the improvement to the mode I delamination resistance is expressed as G_{IR}/G_{Ic}, which is the interlaminar toughness of the stitched composite (G_{IR}) normalised to the toughness of the equivalent unstitched laminate (G_{Ic}). The figure shows a general increase to the interlaminar fracture toughness with increasing stitch density. A few outlying data points show that the delamination resistance can be improved by over 30 times by stitching with exceptionally thick, strong threads. For most composites, however, stitching increases the delamination resistance by a factor of up to 10-15. This compares favourably with other types of 3D composites that have interlaminar fracture toughness properties that are up to 20 times higher than the equivalent 2D laminate.

A number of micromechanical models have been proposed to determine the improvement to the mode I interlaminar fracture toughness properties of composites due to stitching. Of the models, there are two models proposed by Jain and Mai that have proven the most accurate (Jain and Mai, 1994a, 1994b, 1994c). Both models are based on Euler-Bernoulli linear-elastic beam theory applied to a stitched composite with the double cantilever beam (DCB) geometry, as illustrated in Figure 8.22. The models can be used to calculate the effect of various stitching parameters (eg. stitch density, thread strength, thread diameter) on the R-curve behaviour and G_{IR} value of any laminated composite.

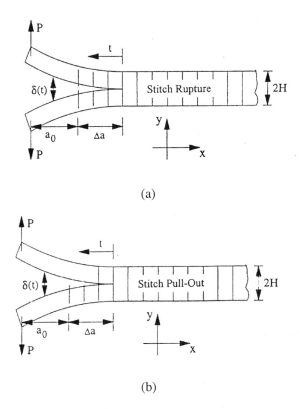

(a)

(b)

Figure 8.22 The DCB specimen geometry used as the basis for the Jain and Mai model for mode I interlaminar fracture toughness of stitched composites. Models have been developed for the cases where the stitches (a) rupture along the delamination crack path (continuous stitching model) and (b) failure at the surface and then pull-out from the composite (discontinuous stitching model) (From Jain and Mai, 1997).

The first model proposed by Jain and Mai is known as the 'continuous stitching model'. With this model it is assumed the stitches are interconnected and fail along the delamination crack plane (Figure 8.22a). This type of failure is also shown in Figure 8.18a. The analytical expression for crack closure traction in the model contains terms for frictional slip and elastic stretching of the stitches in the bridging zone as well as an analytical term to predict when the stitches will rupture at the crack plane. The second model by Jain and Mai is known as the 'discontinuous stitching model'. For this model it is assumed the stitches behave independently under mode I loading, and interlaminar toughening occurs by the frictional resistance of the stitches as they are pulled from the composite under increasing crack opening displacement (Figure 8.22b). To model this failure process the expression for calculating the crack closure traction contains terms for frictional slip and pull-out of the stitches. In some composites, stitch failure occurs during elastic stretching at the outer surface of the DCB specimen at the stitch loop, and the stitch thread subsequently pulls-out. In this case, the continuous and discontinuous stitching models are combined into the so-called 'modified model' to account for the two stitch failure events.

The mode I delamination resistance in terms of stress intensity factor, $K_{IR}(\Delta a)$, of a composite with bridging stitches can be calculated from the expression (Jain and Mai, 1994a, 199b, 1994c):

$$K_{IR}(\Delta a) = K_{Ic} + Y \int_{t=0}^{\Delta a} P(t) \frac{1}{\sqrt{h_c}} f\left(\frac{t}{h_c}\right) dt \tag{8.2}$$

where K_{Ic} is the critical interlaminar fracture toughness of the unstitched composite, Δa is the crack growth length, h_c is the half-thickness of the composite, t is the distance from the crack tip to the specimen end, $P(t)$ is the closure traction due to stitches, and Y and $f(t/h_c)$ are orthotropic and geometric correction factors, respectively. Y is defined by:

$$Y = \sqrt{\frac{E_0}{E_c}} \tag{8.3}$$

where E_0 is the orthotropic modulus and E_c is the flexural modulus of the stitched composite. The term $f(t/h_c)$ in equation 8.2 is determined using:

$$f\left(\frac{t}{h_c}\right) = \sqrt{12}\left(\frac{t}{h_c} + 0.673\right) + \sqrt{\frac{2h_c}{\pi t}} - \left[0.815\left(\frac{t}{h_c}\right)^{0.619} + 0.429\right]^{-1} \tag{8.4}$$

The closure traction, $P(t)$, which is required to determine $K_{IR}(\Delta a)$, is obtained by iteratively solving the Euler-Bernoulli beam equation. Once $K_{IR}(\Delta a)$ has been determined, the Mode I interlaminar fracture toughness, $G_{IR}(\Delta a)$, may be obtained by:

$$G_{IR}(\Delta a) = \frac{K_{IR}^2(\Delta a)}{E_0} \qquad (8.5)$$

The Jain and Mai models have proven reasonably reliable for predicting the delamination properties of stitched composites. For example, Figure 8.23 shows the measured R-curve for a stitched glass/vinyl ester composite (that was shown earlier in Figure 8.15) together with the theoretical R-curve predicted using the Jain and Mai model, and there is good agreement between the two curves. As another example, Figure 8.24 compares the G_{IR} values measured for stitched carbon/epoxy composites against theoretical G_{IR} values calculated using the continuous and modified stitching models. Excellent agreement exists for the modified stitch model while the G_{IR} values are underestimated by about 50% with the continuous model. The accuracy of the models is critically dependent on the failure mode of the stitch, that is whether failure occurs by thread breakage, thread pull-out or a combination of these two.

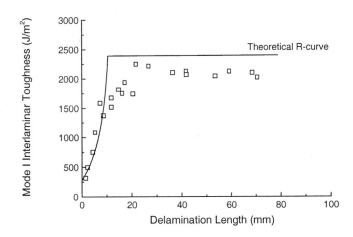

Figure 8.23 Comparison of a theoretical and experimental mode I R-curve for a stitched glass/vinyl ester composite. The theoretical curve was determined using the Jain and Mai model.

8.4.2 Mode II Interlaminar Fracture Toughness Properties

Stitching is also an effective technique for improving the delamination resistance under mode II loading (i.e. shear crack opening). This is particularly significant because delamination cracks that form in composites under impact loading grow mostly under the action of impact-induced shear strains. The effectiveness of stitching in raising the mode II delamination resistance is shown in Figure 8.25, which shows a large increase to the mode II interlaminar fracture toughness (G_{IIR}) of a carbon/epoxy laminate with increasing stitch density (Dransfield et al., 1995). It is worth noting, however, that the improvement to the delamination resistance is usually not as high as for the mode I toughness for equivalent stitch densities. Most stitched composites exhibit a G_{IIR} value that is typically 2 to 6 times higher than the unstitched laminate, depending on the type and amount of stitching. It was shown earlier that the mode I delamination resistance can be increased by much more this.

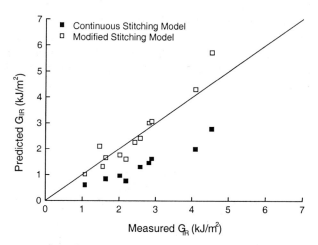

Figure 8.24 Plot of measured against theoretical G_{IR} values for stitched composites. The theoretical G_{IR} values were determined using the modified and continuous stitching models by Jain and Mai. The closer the data points are to the straight line the better the agreement between the measured and theoretical G_{IR} value (Adapted from Mouritz and Jain, 1999).

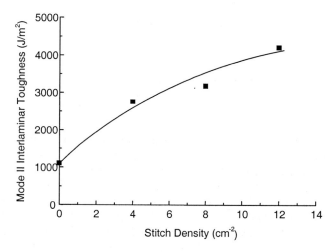

Figure 8.25 The effect of stitch density on the mode II interlaminar fracture toughness of a carbon/epoxy composite (Data from Dransfield et al., 1995).

The toughening mechanisms responsible for the high mode II interlaminar fracture toughness of stitched composites are complex, with a number of different mechanisms operating along the length of a delamination crack. The shear tractions generated in stitches with increasing sliding displacement between the opposing crack faces are shown in Figure 8.26. This figure by Cox (1999) shows typical sliding displacement and stress levels associated with the various mechanisms during shear loading of a stitched composite up to the point of failure. The sliding displacement $(2u_l)$ is the distance the two crack faces have separated under mode II loading. The vertical scales show the average bridging traction across the stitches, τ_b, (left-hand side) and the bridging traction for a single stitch, T (right-hand side). The values shown for τ_b are representative, and will vary depending on the volume fraction of stitching and the mechanical properties of the threads.

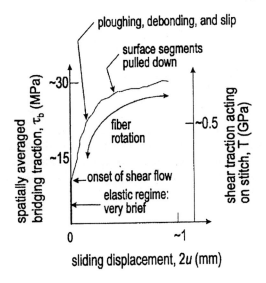

Figure 8.26 Schematic of the shear tractions for mode II loading of a stitch under increasing crack sliding distance (from Cox, 1999).

It is generally acknowledged that when an interlaminar shear stress is applied to a stitched composite containing a delamination then the stitches ahead of the crack front are not damaged or deformed. When the crack tip reaches the stitches, however, the delamination causes the stitches to debond from the surrounding composite material. The stitches are usually completely debonded from the composite when the total sliding displacement $(2u_l)$ exceeds about 0.2 mm. As the opposing crack faces continue to slide pass each other the stitches become permanently deformed. Plastic deformation of the stitches can occur immediately behind the crack tip due to the low shear yield stress of the thread material. It is estimated that permanent deformation in stitches begins when the sliding displacement distance exceeds about 0.1 mm. The stitches experience increasing plastic shear deformation and axial rotation the further they are behind the crack tip. As the stitches are deformed they are ploughed laterally into the crack faces of the composite. At a high amount of axial rotation the stitches experience splitting cracks and spalling, and this generally occurs when the sliding displacement rises above 0.6 mm. This deformation and damage to a sheared stitch is shown in Figure 8.27, and it is obvious a large degree of axial rotation has occurred on the fracture plane. In this thread the fibres have been rotated by an angle (θ) of up to about 45°. The plastic deformation and ploughing of the stitches absorbs a large amount of the applied shear stress. Furthermore, the large amount of axial rotation to the stitches causes them to bend near the fracture plane so a significant load of the applied shear stress is carried by the stitches in tension. The combination of these effects lowers the shear strain acting on the crack tip and thereby improves the delamination resistance. Eventually the stitches at the rear of the stitch bridging zone break when the sliding displacement exceeds about 1 mm (Figure 8.27b). The stitch bridging zone can grow for long distances (up to ~50 mm) before the stitches fail, and this is the principle toughening mechanism against mode II delamination cracks.

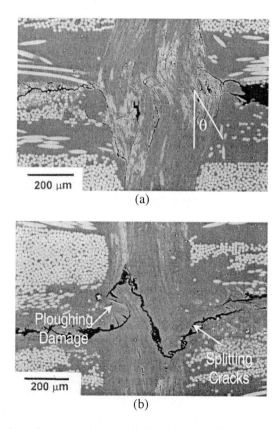

(a)

(b)

Figure 8.27 Scanning electron micrograph showing (a) plastic shear deformation and (b) shear failure to a stitch subject to mode II interlaminar loading.

Micromechanical models have been proposed by Jain and Mai (1994e, 1995) and Cox et al. (Cox, 1999; Cox et al., 1997; Massabò et al., 1998, 1999; Massabò and Cox, 1999) for determining the mode II delamination resistance of stitched composites. The models by Jain and Mai use first order shear deformation laminated plate theory and Griffith's theory for strain energy release rate in fracture to calculate the effect of stitching on the mode II interlaminar fracture toughness (G_{IIR}). Models have been proposed for stitched composites subject to shear loading using the end notched flexure (ENF) and end notched cantilever (ENC) test methods, which are methods for measuring the mode II interlaminar fracture toughness of laminated materials. In both models it is assumed that as a delamination crack propagates under shear the stitch failure process consists of elastic stretching of the threads due to relative slip of the top and bottom sections of the delaminated region, followed by rupture of the stitch in the crack plane. These assumptions do not accurately reflect the actual stitch failure process that has been observed in many stitched composites, which as described above consists of axial plastic shear rotation, splitting/spalling, and ploughing of the stitches.

Jain and Mai (1994e, 1995) state that the mode II strain energy release rate for crack propagation is given by:

$$G_{II} = \frac{A^*}{\cosh^2(\lambda \Delta a)} \left\{ \tau \left(\frac{\sinh(\lambda \Delta a)}{\lambda} + a_0 + \alpha h_c \right) - \frac{\lambda}{A^*} \left(\frac{\alpha_1}{\alpha_2} \right) \sinh(\lambda \Delta a) \right\}^2 \qquad (8.6)$$

where τ is the applied shear stress and is related to the applied load, α is a correction factor accounting for shear deformation, α_1 and α_2 are stitching parameters, and λ is related to materials properties through A^* and α_1. Using the steady-state crack propagation condition, $G_{II} = G_{IIc}$, where G_{IIc} is the mode II critical strain energy release rate for the unstitched composite, the shear stress τ needed for crack propagation can be determined. The critical strain energy release rate for a stitched composite can then be calculated from:

$$G_{IIR} = A^* \tau^2 (a + \alpha h_c)^2 \qquad (8.7)$$

The accuracy of the Jain and Mai models for determining the mode II interlaminar fracture toughness of stitched composites is shown in Figure 8.28. This figure presents a comparison of the measured and theoretical G_{IIR} values for stitched composites, and there is good agreement. However, some studies (eg. Cox, 1999) show significant disagreement between the model and experimental data.

Cox and colleagues have formulated one-dimensional analytical models for predicting the traction shear stress generated in through-thickness fibres (including stitches) when subject to mode II loading (Cox et al., 1997; Cox, 1999; Massabó et al., 1998; Massabó and Cox, 1999). The models are based on the relationship between the bridging tractions applied to the fracture surfaces by the unbroken stitches and the opening (mode I) and sliding (mode II) displacements of the bridged crack. The models consider the micromechanical responses of stitches bridging a delamination crack, including the elastic stretching, fibre rotation and some other affects that occur under mode II. Criteria for failure of the bridging tow by rupture or pull-out is also considered in the models, leading to predictions of the ultimate strength of the bridging ligaments in mixed mode conditions.

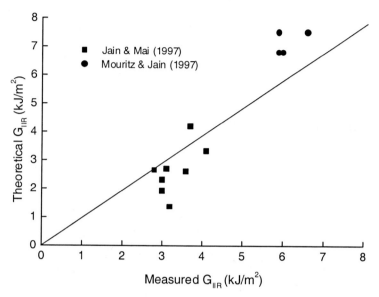

Figure 8.28 Plot of measured against theoretical G_{IIR} values for stitched composites. The theoretical G_{IIR} values were determined using the Jain and Mai models. The closer the data points are to the straight line the better the agreement between the measured and theoretical G_{IIR} value (from Mouritz and Jain, 1999).

Cox (1999) has shown that the bridging shear traction (T_1) generated in a single stitch can be related to the crack sliding displacement (u_1) and crack opening displacement (u_3) by the expressions:

$$u_1 \approx \frac{\sigma_o}{P_x / s}\left[1 - \sqrt{1 - \frac{(T_1 - \tau_o)^2}{\sigma_o^2}}\right] \tag{8.8a}$$

$$u_3 \approx \frac{\sigma_o^2}{2E_t \tau} - \frac{\sigma_o \sin^{-1}\left[\dfrac{T_1 - \tau_o}{\sigma_o}\right] - (T_1 - \tau_o)}{P_x / s} \tag{8.8b}$$

where σ_o is the axial stress in the stitch on the fracture plane, E_t is the Young's modulus of the stitch, τ is the applied shear stress, τ_o is the shear flow stress of the stitch, P_x is the crush strength of the composite, and s is the circumferential length of the stitch. The build-up in the traction stress within a stitch with increasing sliding displacement can be accurately predicted using the above equation. For example, Figure 8.29 compares the predicted traction stress (thick line) with the experimentally measured traction stresses (the two thinner curves) generated in a single Kevlar stitch subject to increasing sliding displacement. The theoretical curve was calculated using the above equations by Cox (1999) and the experimental curves were measured by Turrettini (1996). There is excellent agreement between the theoretical traction curve and the two experimental curves up to the peak stress ($T_1 \sim 1000$ MPa), at which point failure of the stitch occurs. By determining the traction stress generated in a single stitch, it is then possible to determine the average traction stress (t) in a number of stitches bridging a mode II delamination crack in a composite using the simple expression (Cox, 1999):

$$t = c_t T \tag{8.9}$$

where c_t is the area fraction of stitching.

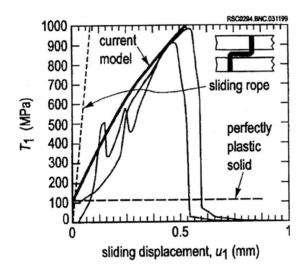

Figure 8.29 Comparison of the Cox model for the shear traction in a single stitch (thick curve) with two experimental curves showing measured traction in a Kevlar stitch in a carbon/epoxy laminate determined by Turrenttini (1996) (from Cox, 1999).

8.5 IMPACT DAMAGE TOLERANCE OF STITCHED COMPOSITES

8.5.1 Low Energy Impact Damage Tolerance

As discussed in Chapter 1, a problem with using 2D laminated composites in highly-loaded structures, particularly aircraft components, is their susceptibility to low energy impact damage. The damage caused by a low energy impact is characterised by delamination cracking, matrix cracking and, in some instances, breakage of fibres. Low energy damage to thin aircraft grade composites usually occurs at incident impact energies between 1 and 5 J. The delaminations caused by an impact can reduce the strength, particularly under compression loading, and thereby degrades the structural integrity of composite components. A key strategy to improve the impact damage tolerance of composites is to provide through-thickness reinforcement against delamination cracking using stitching. As described in Section 8.4, stitching is highly effective in improving the interlaminar fracture toughness of laminated composites, and therefore it is expected that stitched materials will have a high resistance to delamination cracking under impact loading.

The effectiveness of stitching in suppressing low energy impact damage has been thoroughly investigated for a variety of FRP composites, including carbon/epoxy, and most stitched materials respond in a similar way to impact loading (Bibo and Hogg, 1996; Caneva, 1993; Cholakara et al., 1989; Dow and Smith, 1989; Farley et al., 1992; Funk et al., 1985; Liu, 1987; Liu, 1990; Mouritz et al., 1996b; Ogo, 1987; Pelstring and Madan, 1989; Sharma and Sankar, 1994; Wu and Liau, 1994; Wu and Wang, 1994). It appears that the effectiveness of stitching is critically dependent on the length the delaminations have spread from the impact site. Stitching does not usually increase the threshold impact energy needed to form and initiate the growth of delaminations. This is because it does not raise the strain energy needed to initiate delamination cracks.

The effectiveness of stitching in improving the damage resistance of composites is critically dependent on the incident impact energy. Stitching does not usually improve the damage resistance when the energy impact is low (Herszberg et al., 1996; Leong et al., 1995; Leong et al., 1996; Mouritz et al., 1996). This behaviour is shown in Figure 8.30 which compares the amount of damage to stitched and unstitched composites caused by low energy impacts. This figure shows the amount of damage to the stitched and unstitched materials is similar over the range of impact energies. The inability of stitching to improve the damage resistance is probably due to the short length of the delamination cracks. When the impact energy is low then the delaminations rarely grow longer than 10-20 mm before stopping. In Section 8.4 it was shown that the ability of stitching to suppress delamination cracking is small for short cracks because the stitch bridging zone is not fully developed. As a result, stitching is not highly effective in reducing the amount of damage when the delaminations formed by an impact are short. Under these impact conditions, the post-impact mechanical properties, such as compression-after-impact strength, of stitched composites are similar or marginally lower than the equivalent unstitched material (Herszberg et al., 1996; Leong et al., 1995; Leong et al., 1996; Mouritz et al., 1996).

Stitching is highly effective in suppressing delamination damage at medium-to-high impact energies. The ability of stitching to improve the damage resistance appears to become increasingly effective when the incident impact energy exceeds about 3 to 5 J/mm. An example of the improved impact damage resistance that can be achieved with stitching is shown in Figure 8.31 (Wu and Liau, 1994). This figure compares the length of delamination cracks in stitched glass/epoxy composites against the equivalent unstitched laminate. It is seen that the amount of damage is reduced by stitching when the impact energy exceeds ~2 J/mm. The effectiveness of stitching in reducing the amount of damage then becomes more pronounced with increasing impact energy. At relatively high impact energies, long delaminations are formed which allows the full development of a stitch bridging zone. As a result, the stitched materials are highly effective in reducing the extent of delamination damage caused by an impact.

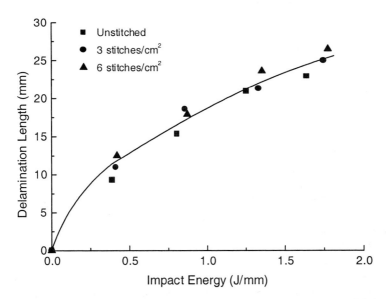

Figure 8.30 Effect of very low energy impact loading on the amount of delamination damage caused to an unstitched glass/vinyl ester composite and the same material stitched with Kevlar yarn.

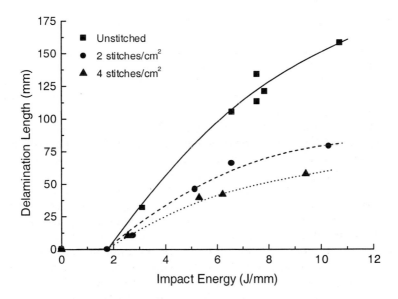

Figure 8.31 Effect of low energy impact loading on the amount of delamination damage caused to stitched and unstitched composites (Data from Wu and Liau, 1995).

The ability of stitching to reduce the amount of damage improves not only with the incident impact energy. The effectiveness of stitching also improves dramatically with stitching density, as shown in Figure 8.32 (Liu 1990). In the figure the normalised delamination area defines the amount of impact damage to the stitched composite divided by the amount of damage to the equivalent unstitched laminate. There is a rapid reduction to the amount of impact damage with increasing stitch density, and in this case it is seen that stitching reduced the delamination area by as much as 40% compared with the unstitched laminate.

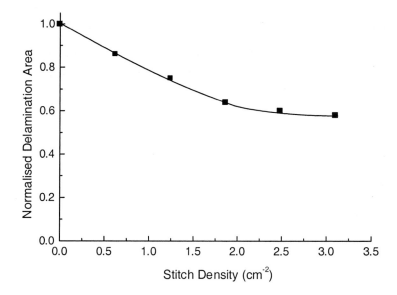

Figure 8.32 Effect of stitch density on the amount of impact damage to a glass/epoxy composite. The composite was impacted at an energy of about 7.5 J/mm (Data from Liu, 1990).

The improved damage resistance provides stitched composites with higher post-impact mechanical properties than the unstitched material. For example, Figure 8.33 (Rossi, 1989) compares the compression-after-impact strengths of a stitched and unstitched carbon/thermoplastic composite. It is seen the compression-after-impact strength of the stitched composite is slightly higher. The higher post-impact strength is attributed to two factors: firstly, the amount of delamination damage in lower in the stitched material, and secondly, the stitches suppress the growth of the delaminations and inhibit sublaminate buckling under compression loading.

Models for estimating the compression-after-impact strength of stitched composites have not yet been formulated because of the complexity of modeling the growth of multiple delaminations and the subsequent multiple sublaminate buckling processes that can occur under compression. However, models have been developed for predicting the compression strength of stitched laminates containing a single delamination (Shu and Mai, 1993a, 1993b). These models provide insights into the effectiveness of stitching in improving the compression-after-impact strength. A model proposed by Cox (2000) states that the critical uniaxial compressive stress needed to induce sublaminate buckling within a stitched composite containing a single delamination can be expressed by:

$$\sigma_b^\infty = -\frac{5}{3\sqrt{3}}\sqrt{c_s E_s E_1}\sqrt{\frac{h}{t}}$$

(8.10)

where c_s is the area fraction of stitches, E_s is the Youngs modulus of the stitches, E_1 is the Youngs modulus of the composite in the load direction, h is the thickness of the delaminated layer, and t is the thickness of the entire laminate. This equation shows that the buckling stress increases with the area fraction of stitching, and this explains why stitched composites usually have higher compression-after-impact strengths than the unstitched laminate. Equation 8.10 also reveals that the compression-after-impact strength can be improved by using stitches having a high modulus.

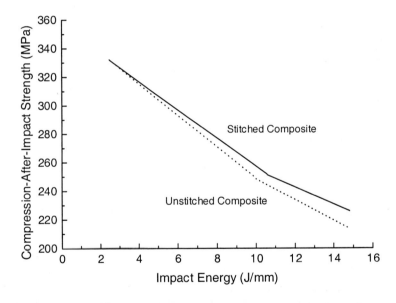

Figure 8.33 Effect of impact energy on the compression-after-impact strengths of a stitched and unstitched carbon/thermoplastic composite (Data from Rossi, 1989).

8.5.2 Ballistic Impact Damage Tolerance

The potential use of stitched composites in military aircraft and helicopters has prompted an assessment of their impact damage tolerance to ballistic projectiles such as bullets (Kan and Lee, 1994; Keith, 1999; Mouritz, 2001). Ballistic projectiles travel at velocities between 450 and 1250 m/s and easily perforate thin composite laminates and cause extensive delamination damage around the bullet hole. Stitching has proven effective in reducing the amount of delamination damage caused by a ballistic projectile, resulting in higher post-impact mechanical properties than the unstitched laminate. The effect of the amount of stitching on the compression-after-ballistic impact strength of a carbon/epoxy composite is shown in Figure 8.34. The strength values shown were determined after a tumbling 12.7 mm projectile travelling at high speed had perforated the composite. The post-impact strength is seen to rise steadily with the volume percent of stitching, and this clearly demonstrates that stitching is an effective technique in improving the ballistic impact damage tolerance of composite materials.

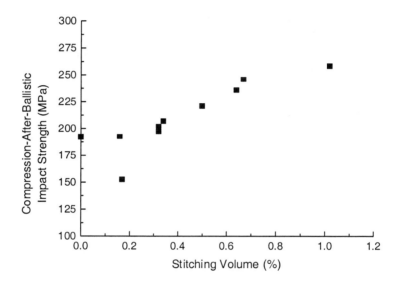

Figure 8.34 Effect of stitching content on the compression-after-ballistic impact strength of a carbon/epoxy composite (Data from Keith, 1999).

8.5.3 Blast Damage Tolerance

The potential use of stitched composites in military structures has led to an evaluation of their damage tolerance to explosive blasts (Mouritz 1995a, 1995b, 2001). Blast studies have revealed that stitching is highly effective in reducing the amount of delamination damage caused by the shock wave from an explosion. For example, Figure 8.35 shows the effect of stitch density on the amount of blast damage and the flexure-after-blast strength of a composite (Mouritz, 2001). The results shown are for the composite subject to a medium and high intensity explosive blast. It is seen that the amount of delamination damage decreases with increasing stitch density, and this results in the stitched composites having similar or higher post-blast flexural strengths than the unstitched laminate. The superior ballistic and explosive blast damage tolerance properties of stitched composites indicate that these materials are ideally suited for use in military aircraft.

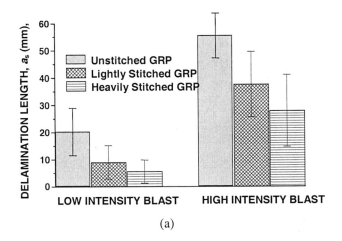

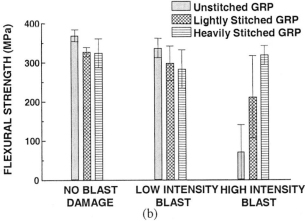

Figure 8.35 (a) Amount of delamination damage caused by a low and high intensity explosive blast. (b) Flexure-after-blast strengths of stitched and unstitched composites (Mouritz, 2001).

8.6 STITCHED COMPOSITE JOINTS

For adhesively bonded composite lap joints, typical failure initiates and propagates, in a form of delamination, along the interface between the surface and the second ply in one composite adherend. Figure 8.36 schematically depicts the onset and propagation of interlaminar delamination between the surface and second plies in a double-lap composite joint. It is believed that the high positive normal stress near an overlap end and the low interlaminar strength are believed to be the two major contributing factors. Depending on the joint configuration and loading conditions, a delamination can propagate along an interface or kink into an adjacent interface, or a sectional fracture occurs in the deformed surface ply.

The strength of typical composite lap joints can be limited by the interlaminar strength, which is the weak link for composite adherends as it relies on the brittle matrix tensile properties and the bonding strength of the fiber/matrix interface. To improve composite lap joint strength, one can choose a toughened resin system for the composite substrate to increase the interlaminar fracture toughness and/or taper the composite substrate in a form of ply drop-off to reduce the positive normal stress.

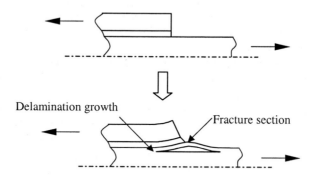

Figure 8.36 Peel stress induced interlaminar delamination in composite lap joints

Placement of fibres in the through-thickness direction using the stitching and z-pinning technique provides a bridging mechanism holding the two delaminated substrates together. Sawyer (1985) utilized prepreg to laminate the composite substrates in single-lap joints, which were then transversely stitched using a shoe-making sewing machine. Comparison of the failure loads of the joints with and without transverse stitching revealed that transverse stitching can significantly improve the static strength of the joints.

Instead of stitching the prepreg, which causes appreciable fibre damage, Tong et al (1998) stitched dry fabric preform, which was then placed in a mould and resin was injected using the resin transfer moulding technique, to demonstrate the promising effect of transverse stitching. Figures 8.37 and 8.38 illustrate the configurations of the single-lap joint specimen and the stitching pattern.

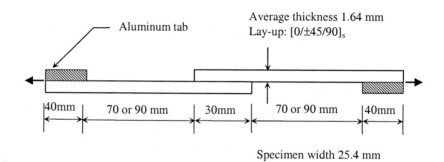

Figure 8.37 Configuration of the single-lap joint specimen manufactured using the RTM process

In the experiments performed by Tong et al (1998), the specimens were prepared by (a) overlaying two [0/±45/90]$_s$ fabric stacks followed by debulking under vacuum and heat to produce a preform of single-lap panel; (b) applying transverse stitches following the designed pattern; and (c) injecting resin and consolidating the panel under clamping pressure and a curing temperature of 80°C for 4 hours.

All specimens were manufactured from Ciba Composites Injectex® uniweave carbon fabric GU230-E01 and GY260 epoxy resin/HY917 hardener/DY070 accelerator. The uniweave material has 90% of its fibers oriented in the warp direction and the remaining fibers in the weft direction to hold the warp fibres in place for ease of handling. The Injectex® has been developed for precise fabric placement at preform stage prior to resin infusion. The stitch material used was a twisted 40tex (2×20) Kevlar thread, and zigzag stitching pattern was employed with the overstitch limited to 1 mm as schematically shown in Figure 8.38.

The measured axial loads increased almost linearly with the applied axial displacement for all specimens up to the final failure. For all specimens catastrophic failure occurs upon attaining the ultimate load. The average failure loads are tabulated in Table 8.1. The results show that the stitched single-lap joints are stronger than the unstitched ones. For the long specimens with an unsupported length of 90 mm, through-thickness stitching leads to an average increase in joint strength by about 25%. For the short specimens with an unsupported length of 70 mm, there is an average increase in joint strength of about 22%.

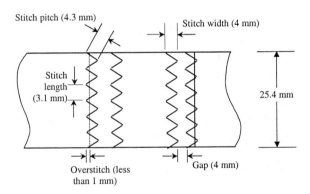

Figure 8.38 Top view of the four-row zigzag stitch used in the overlap of the single-lap joint (Tong et al, 1998a)

Table 8.1 Effect of stitching on static failure strength of single lap joints fabricated by stitching preform and using RTM

Specimen group	Unsupported length (mm)	Average failure load (kN)
Unstitched	90 mm	11.33
Unstitched	70 mm	12.37
Stitched	90 mm	14.11
Stitched	70 mm	15.06

Figure 8.39 plots the applied load versus the number of cycles to failure for the stitched and unstitched specimens subjected to a tension-tension load of R=5 at a frequency of 3 Hz. The specimens are tested to failure or up to 10^6 cycles. Clearly, transverse stitching can improve the fatigue life by two orders of magnitudes for any given maximum tensile load. For a given cycle life, stitched specimens carry a significantly higher load than the unstitched specimens. In addition, for stitched joints, stable crack propagation along the interface between the two adherends is observed when the maximum load is only a fraction of the static strength of the unstitched specimens. The through-thickness stitches are found to bridge the cracked specimens.

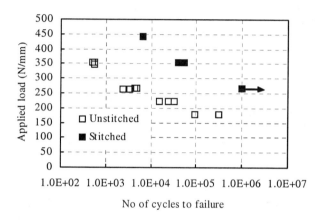

Figure 8.39 Effect of stitching on fatigue strength of single–lap joints loaded with a load ratio R=5 and at a frequency of 3 Hz

Tong et al (1998) also performed another set of experimental tests of single-lap joint specimen manufactured following the RTM process from the Hexcel Composites G926 EFP, 6k, 5 Harness satin weave carbon fabric and Ciba Araldite LY 556/ Hardener HY 917/ Accelerator DY 070 epoxy resin. All specimens had an overlap length of 30 mm, an unsupported length of 60 mm, and a width of 25.4 mm. The adherend had a lay-up of [0/-45/45/90]$_s$, and a nominal thickness of 2.864 mm. Kevlar 40 tex thread was used as the stitching thread. A zigzag stitch pattern was used at the overlap ends and a straight plain stitch pattern was used in the central region of the overlap with modified interlocking stitch. It was found that the transverse stitching can improve the average failure load by 41%.

Chapter 9

Z-Pinned Composites

9.1 INTRODUCTION

The technology of reinforcing composites in the through-thickness direction with small pins was first evaluated in the 1970s. Thin steel pin wires were inserted at offset angles of ±45° into carbon/epoxy prepreg laminates to improve the delamination toughness (Huang et al., 1978). The pins used were very thin, with a diameter of only 0.25 mm, to minimise damage to the laminates. The steel pins were effective in increasing the interlaminar shear strength and delamination resistance. However, initially it was neither practical nor cost-effective to insert thin pins over a large area of composite material, and therefore the technology was not immediately taken-up by the aircraft composites industry.

Z-pinning technology was developed further in the early 1990s by Aztex Inc. The technology involves embedding small diameter pins, known as Z-fibers[TM], into composites to produce a 3D fibre network structure, as illustrated in Figure 9.1. Z-fiber[TM] technology is the newest of the various techniques for producing 3D composites, and already it has a wide variety of potential applications in engineering structures. An important potential use of Z-fibers[TM] is for the attachment and reinforcement of composite joint structures such as lap joints, T-joints and rib stiffeners. Z-pins are being used to fasten hat-stiffened sections to the composite skins in selected parts of the F/A-18 Hornet fighter aircraft. Z-fibers[TM] can be used in composite joints in place of bolted fasteners or rivets to provide a more evenly distributed load over the joint area. Z-fibers[TM] can also be used for the local reinforcement of composite panels to reduce the incidence of edge delaminations as well as the reinforcement of sandwich panels to minimise the likelihood of skin peeling and debonding.

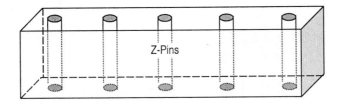

Figure 9.1 Schematic illustration of a z-pinned composite

The relatively recent development of Z-fibers[TM] has meant that z-pinned composites have not been explored in detail. In this chapter the current state of knowledge of z-pinned composites is examined based on the limited amount of published data and information. Included in this chapter is a description of the fabrication techniques, the in-plane and through-thickness mechanical properties, and impact damage tolerance of z-pinned composites. 3D sandwich composites manufactured using Z-fiber™ technology is also briefly described.

9.2 FABRICATION OF Z-PINNED COMPOSITES

The manufacture of 3D composites with Z-fibers™ is a multi-stage process that can be performed inside an autoclave or in a workshop using an ultrasonic tool. The fabrication of z-pinned composites is similar to the manufacture of stitched composites in that the through-thickness reinforcement is inserted as a separate processing step to create a 3D composite material. In this way these reinforcement techniques are different to the textile technologies of weaving, braiding and knitting that create an integrated 3D fibre preform within a single stage process. The steps involved in z-pinning using the autoclave process are shown schematically in Figure 9.2.

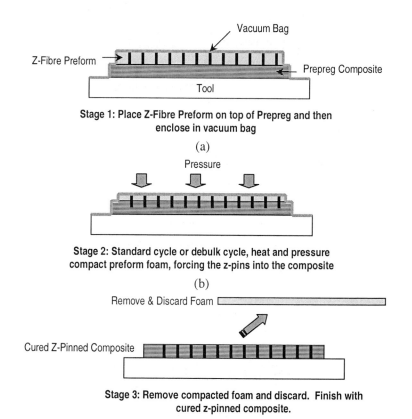

Stage 1: Place Z-Fibre Preform on top of Prepreg and then enclose in vacuum bag

(a)

Stage 2: Standard cycle or debulk cycle, heat and pressure compact preform foam, forcing the z-pins into the composite

(b)

Stage 3: Remove compacted foam and discard. Finish with cured z-pinned composite.

(c)

Figure 9.2 Schematic of the Z-fiber™ insertion process using an autoclave (Adapted from Freitas et al., 1994)

The process begins by overlying an uncured composite prepreg laminate with an elastic foam preform containing a grid of Z-fibers™. The purpose of the foam is to keep the rods in a vertical position and to stop them from buckling as they are embedded into the composite. Once the preform is in position it is compressed under pressure from the autoclave which forces the pins into the underlying composite (shown as step 2 in Figure 9.2). The pins are chamfered at one end to an angle of about 45° to easily pierce the composite. The benefit of inserting the pins within an autoclave is that the composite can be heated to reduce the viscosity of the resin matrix. This allows the Z-fibers™ to penetrate more easily and thereby reduce the amount of fibre damage to the composite. After the pins are embedded in the composite the residual foam is removed and the fabrication process is complete (step 3 in Figure 9.2). Any excess pin material that is protruding above the composite surface can be easily removed using a shear cutting tool.

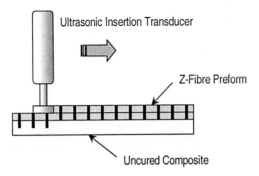

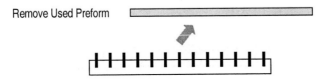

(a) Primary Insertion Stage & Residual Preform Removal

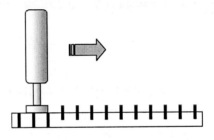

(b) Secondary Insertion Stage

Figure 9.3 Schematic of the Z-fibre™ insertion process using UAZ. (Adapted from Freitas et al., 1996)

Alternatively, the pins can be inserted into uncured prepreg laminates without an autoclave using a specially designed ultrasonic tool, as shown in Figure 9.3. This method of pinning is known as the 'Ultrasonically Assisted Z-Fiber[TM] (UAZ) process'. In the UAZ process the foam preform is partially compacted under the pressure of high frequency acoustic waves generated by the ultrasonic tool. This forces the pins partway into the composite, and at this stage the residual foam is removed and a second pass is made with the ultrasonic tool to drive the pins completely into the composite.

Z-fibers[TM] can be inserted into most types of FRP materials, including uncured tape laminates, pre-consolidated thermoplastic composites, cured thermoset composites, and dry fabric preforms. (Z-fibers[TM] can also be used to reinforce and bond aluminium sheet laminates (Freitas and Dubberly, 1997)). Z-fibers[TM] are made of protruded composite material or metal rod and range between 0.15 and 1.0 mm in diameter. Composite materials used in Z-fibers[TM] include high-strength carbon/epoxy, high-modulus carbon/epoxy, carbon/bismaleimide (BMI), S-glass/epoxy, and silicon carbide/BMI. The metal pins are made of titanium alloy, stainless steel and aluminium alloy. The amount of Z-fibers[TM] used for pinning composites is normally in the range of 0.5% to 5% of the total fibre content, although it is possible to have a greater amount for high through-thickness reinforcement.

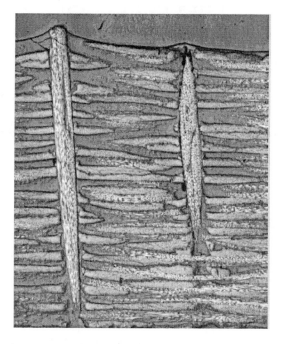

Figure 9.4 Z-pins in a composite (Courtesy of the Cooperative Research Centre for Advanced Composite Structures Ltd)

The fibre structure of a composite reinforced with Z-fibers[TM] is shown in Figure 9.4. The pins extend through-the-thickness of the composite, and are usually inclined at a small angle (less than ~7°). The damage caused to composites by the insertion of Z-fibers[TM] is still under investigation, and more research is needed to understand the types and amounts of damage caused in the z-pinning process. The limited amount of information that has been published on damage reveals that the most common types are misalignment and fracture of in-plane fibres (Steeves and Fleck, 1999a). The in-plane fibres of unidirectional tape laminates are misaligned and distorted when pushed aside during insertion of Z-fibers[TM]. An example of the distortion of fibres around a z-pin is shown in Figure 9.5. The misalignment angle of in-plane fibres around the pins is dependent on a number of factors, including the type of composite (i.e. prepreg, thermoplastic, fabric preform), fibre lay-up orientations, and the fibre volume content of the laminate. Misalignment angles of between 5° and 15° have been measured in unidirectional tape composites reinforced with Z-fibers[TM], compared to unreinforced tape laminates with an average misalignment angle of about 3°. In some cases the fibres can be so severely misaligned that they break. The incidence of fibre breakage is not known, although it is expected to be small. Due to the misalignment of in-plane tows, resin rich regions are formed at each side of the pin, as shown in Figure 9.5. These regions can be up to ~1 mm long, and in composites reinforced with a high density of Z-fibers[TM] these regions may join up to form continuous channels of resin.

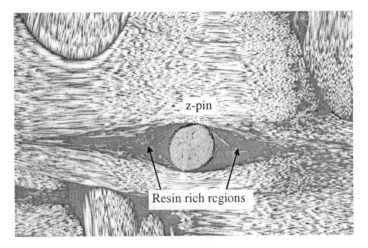

Figure 9.5 In-plane fibre distortion around a z-pin in a composite. Note the presence of the large resin-rich zones

9.3 MECHANICAL PROPERTIES OF Z-PINNED COMPOSITES

The in-plane mechanical properties and failure mechanisms of composites reinforced with Z-fibers[TM] have not been studied in detail. The effect of Z-fiber[TM] reinforcement on in-plane properties such as flexural modulus and strength, shear strength, fatigue-life, and open-hole tensile and compressive strengths have not been reported, and it is an important topic of future research to determine these mechanical properties and failure mechanisms. The discussion here is confined to describing changes to the tensile and compressive strengths of z-pinned composites that is based on a small number of studies. Freitas et al. (1991, 1994, 1996) have shown that the tensile properties of a tape carbon/epoxy laminate are unaffected when the amount of Z-fibers[TM] is low (under 1.5%). However, the tensile properties can be degraded with high amounts of pinning. For example, Figure 9.6 shows the effect of Z-fiber[TM] content on the tensile strength of a unidirectional tape laminate. The strength decreases rapidly as the Z-fiber[TM] content is increased up to an areal density of 10%. It is seen that the loss in tensile strength can be significant, with the strength of the composite with a Z-fiber[TM] content of 10% being only 60% of the original strength. The failure mechanism of z-pinned composites under tensile loading has not been studied. It is speculated, however, that the reduction to tensile strength is due to the misalignment and fracture of fibres. However, more experimental work is needed to identify the tensile failure mechanism and micro-mechanical models are needed for predicting the elastic modulus and tensile strength of z-pinned composites.

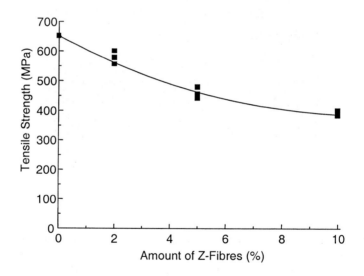

Figure 9.6 Effect of Z-fiber[TM] content on the tensile strength of a carbon/epoxy composite. Adapted from Freitas et al. (1996)

The effect of z-pinning on the compressive properties and failure mechanisms of composites is currently under investigation, and data on the compressive strength of z-pinned composites is limited. Stevens and Fleck (1999a, 1999b) have shown that the compressive properties of composites can be degraded by Z-fibers[TM], with reductions in compressive strength of up to 33% being measured for unidirectional tape laminates. The reduction to compression strength is due to the misalignment of in-plane fibres around the z-pins (as shown in Figure 9.5) that causes the tows to fail by kinking at lower compressive loads. As described earlier in Chapter 5, the failure mechanism of kinking is highly dependent on the degree of misalignment of the in-plane tows. Kinking is initiated in localised regions within a composite where the fibres are misaligned with respect to the applied compressive load by some processing defect or rigid inclusion, such as a z-pin. Under axial compressive loading, shear strains develop in the misaligned tow that promotes localised micro-cracking and crazing of the resin matrix. This damage weakens the tow, which permits further rotation of the fibres until a kink band is formed. Figure 9.7 shows a kink band that has initiated in a composite reinforced with Z-fibers^{TM}. In the absence of a notch, lateral propagation of the kink band across the entire composite is unstable and so failure occurs suddenly.

Figure 9.7 A kink band next to a z-pin in a carbon/epoxy composite (from Steeves and Fleck, 1999a).

The compressive stress required to induce failure by kinking decreases rapidly with increasing initial misalignment angle of the in-planes fibres. Figure 9.8 shows the effect of initial misalignment angle of fibres around a z-pin on the compressive strength of a unidirectional carbon/epoxy tape laminate (Steeves and Fleck, 1999a). The curve shows the theoretical reduction to compressive strength while the two data points show measured strength values for the composite where the initial misalignment angle around the Z-fibers^{TM} was known. This figure reveals that the key factor limiting the compressive strength of z-pinned composites is the amount of fibre distortion caused by the Z-fibers^{TM}.

9.4 DELAMINATION RESISTANCE AND DAMAGE TOLERANCE OF Z-PINNED COMPOSITES

The amount of information on the delamination resistance and impact damage tolerance of z-pinned composites is limited, although the data that has been published shows that these materials are highly resistant to interlaminar cracking and through-thickness failure. Z-fibers[TM] are highly effective in improving the mode I interlaminar fracture toughness (G_{Ic}) of tape laminates (Freitas et al., 1994; Cartié and Partridge, 1999a, 1999b; Grafticaux et al., 2000; Partridge and Cartié, 2001). This is shown in Figure 9.9 that shows the effect of Z-fiber[TM] content on the delamination resistance of a carbon/epoxy laminate. The mode I interlaminar fracture toughness is seen to increase rapidly with the amount of Z-fibers[TM]. The maximum G_{Ic} value of 11.6 kJ/m^2 was achieved with the relatively modest Z-fiber[TM] content of 1.5%, and this level of interlaminar toughness is similar or greater than that achieved by 3D weaving, braiding, knitting or stitching. Cartié and Partridge (1999b) found that the level of interlaminar toughening is also dependent on the diameter of the z-pins, with thinner pins providing higher delamination resistance. For example, it was found that the mode I interlaminar toughness of a carbon/epoxy tape laminate reinforced with 2% Z-fibers[TM] was more than doubled when the pin diameter was reduced from 0.50 mm to 0.28 mm.

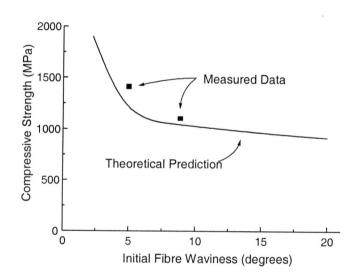

Figure 9.8 Effect of maximum fibre misalignment angle caused by z-pin reinforcement on the compressive strength of a unidirectional carbon/epoxy composite. The curve shows a theoretical prediction and the points are experimentally measured values (Adapted from Steeves and Fleck, 1999a).

Z-pinning is also highly effective in raising the interlaminar fracture toughness of composites under mode II and combined modes I/II loading (Cartié and Partridge, 1999a, 1999b; Partridge and Cartié, 2001). For example, Figure 9.10 shows the effect of combined modes I and II loading on the delamination resistance of a carbon/epoxy composite reinforced with 1% or 2% Z-fibers[TM]. A significant increase in the interlaminar toughness is achieved, particularly with the higher amount of reinforcement.

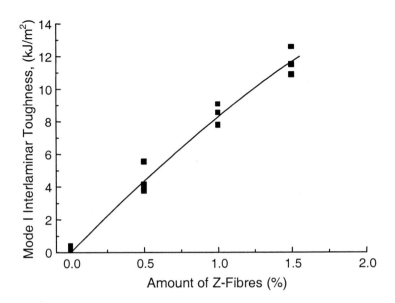

Figure 9.9 Effect of z-pin content on the mode I interlaminar fracture toughness of a carbon/epoxy composite (Data from Freitas et al., 1994)

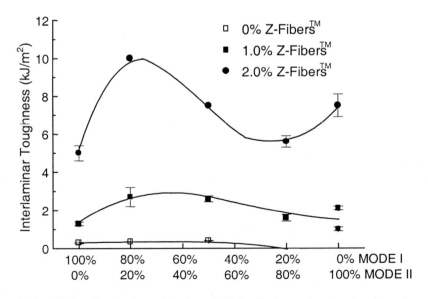

Figure 9.10 Effect of mixed modes I and II loading on the interlaminar fracture toughness of z-pinned carbon/epoxy composites (Data from Cartié and Partridge, 1999a)

It is evident that z-pinning is highly effective in raising the delamination resistance of composites, however further work is required to optimise the pinning conditions to achieve the maximum improvement in interlaminar fracture toughness. The effects of the stiffness, strength, diameter and type of the pin as well as the areal density of pinning on the modes I and II interlaminar fracture toughness needs to be thoroughly investigated. This investigation can be facilitated with the recent development of micro-mechanical models for predicting the interlaminar fracture toughness of z-pinned composites. A model has been proposed by Liu and Mai (2001) for mode I toughness whereas the model by Cox (1999) described in Section 8.4.2 can be used for mode II toughness.

The mechanism of interlaminar toughening that occurs with z-pinned composites is similar to that operating with other types of 3D composites. The pins appear to do little to prevent the initiation of interlaminar cracks, which might be considered as the onset of delamination growth up to a length of about 1 to 5 mm. With cracks longer than about 5 mm, however, the z-pins slow or totally suppress the further growth of delaminations by a crack bridging action. The toughening processes for modes I and II loading are shown schematically in Figure 9.11. Interlaminar toughening occurs by the z-pins bridging the delamination behind the crack front, and through this action is able to support a significant amount of the applied stress. This greatly reduces the strain acting on the crack tip and thereby increases the interlaminar fracture toughness and stabilises the crack growth process. When the separation distance between the crack faces becomes large the rods are pulled from the composite or break. In the case of mode II loading, the pins also absorb a significant amount of strain energy by shear deformation until failure occurs by pull-out or rupture.

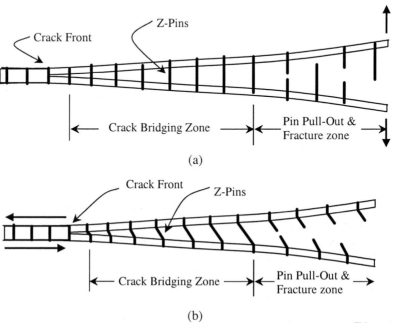

Figure 9.11 Schematic of the bridging toughening mechanism in Z-fiber[TM] composites for (a) mode I and (b) mode II loading

The high interlaminar toughness of Z-fiber[TM] composites makes these materials highly resistant to edge delaminations and impact damage (Freitas et al., 1994). Edge delaminations can be a major problem in composite structures, particularly at free edges and bolt holes, and it is often necessary to taper the edges and reinforce holes to reduce the incidence of cracking. Freitas et al. (1994) found that the tensile load needed to produce edge delaminations in a carbon/epoxy tape composite is increased dramatically with a small amount of Z-fiber[TM] reinforcement. Figure 9.12 shows the tensile load at which the onset of edge delamination cracking occurred in the composite with a Z-fiber[TM] content of 0%, 0.5% or 1.0%. It is seen that the delamination load increased by over 70% when the composite was reinforced with a Z-fiber[TM] content of only 0.5%, and increased by over 90% with 1.0% Z-fibres[TM]. Z-fibers[TM] are also highly effective increasing the fatigue life and stabilising the growth of fatigue damage in blade-stiffened panels (Owsley, 2001).

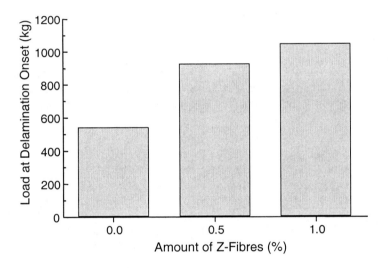

Figure 9.12 Effect of z-pin content on the tensile load needed to induce edge delaminations in a carbon/epoxy composite (adapted from Freitas et al., 1994)

The delamination resistance of composites under impact loading is also improved with z-pinning, although the impact damage resistance of these materials has not been as extensively studied as for other types of 3D composites. A preliminary study has shown that the amount of impact damage experienced by carbon/epoxy composites is reduced by between 30% and 50% with z-pinning (Freitas et al., 1994), and even larger reductions can be expected with a high amount of reinforcement. The improved impact damage resistance provides Z-fiber[TM] composites with higher post-impact mechanical properties than equivalent materials without through-thickness reinforcement. For example, Freitas et al. (1994) found that the compression-after-impact strength of a carbon/epoxy composite was nearly 50% higher when reinforced with Z-fibers[TM].

9.5 Z-PINNED JOINTS

As mentioned earlier, a potential application of Z-fiber[TM] technology is for the reinforcement of composite structures such as lap and blade-stiffened composite joints. While the mechanical performance of z-pinned joints is still being evaluated, preliminary studies reveal that these joints have much better load-bearing properties than joints made using conventional joining techniques such as adhesive bonding or co-curing. For example, Rugg et al. (1998) found that Z-fibers[TM] are highly effective in improving the mechanical properties of composite lap joints, with z-pinning producing a large increase in the shear modulus and about a 100% rise in the failure load.

Freitas et al. (1996) examined the effectiveness of z-pinning in increasing the tensile pull-off strength of a carbon/epoxy T-shaped joint. In the study the tensile pull-off response of an unreinforced T-joint made by co-curing was compared to the joint reinforced with mechanical fasteners, 2% Z-fibers[TM] or 5% Z-fibers[TM]. The profile of a z-pinned joint is shown in Figure 9.13, and the entire joint area was reinforced with z-pins made of titanium alloy, including the stiffener, stiffener radius, and flange ends. The results of the tensile pull-off tests are shown in Figure 9.14. In this figure Curve A and Curve B refer to the unreinforced and bolted joints, respectively, whereas Curves C and D refer to the joint reinforced with 2% and 5% z-pins. All four joints had the same initial failure load of 1600-2000 N, suggesting that z-pins are not effective in suppressing the onset of tensile damage in stiffened panels. However, at this load the unreinforced joint failed catastrophically whereas the reinforced joints continued to support an increasing load. It is seen in Figure 9.14 that the z-pinned joints failed at a tensile load that was 2.3 or 2.6 times higher and a displacement that was 7 or 8 times greater than the unreinforced joint. Furthermore, the z-pinned joints were able to support 70% more load than the joint reinforced with mechanical fasteners.

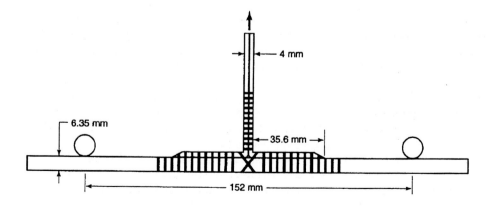

Figure 9.13 Profile of the blade stiffened joint showing the locations of the z-pin reinforcement (Freitas et al., 1996)

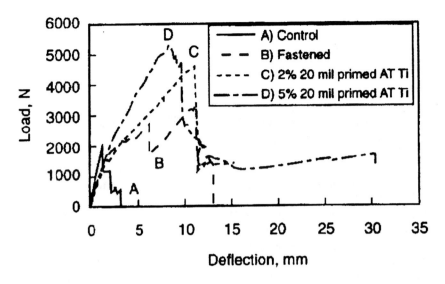

Figure 9.14 Tensile pull-off results of stiffened joints (Freitas et al., 1996)

9.6 Z-PINNED SANDWICH COMPOSITES

Conventional sandwich composite materials used in aircraft, marine craft and civil structures are prone to delamination cracking and failure at the edge of the face skins when subject to high peel stresses. Various techniques have been developed to improve the peel resistance of sandwich composites, including tapered ends to the face skins, bolting or riveting of the face skins, and using high toughness adhesives between the face skins and core. More recently, Z-fiber™ technology has been used to increase the peel strength and provide through-thickness reinforcement to sandwich composites. Aztex Inc., the manufactures of Z-fibers™, produce two products known as X-Cor™ and K-Cor™ which are structural sandwich materials reinforced with z-pins. The z-pins are inserted through the entire thickness of sandwich materials during processing in an autoclave. The process is similar to that shown in Figure 9.2 for the insertion of z-pins into single-skin composites. During processing the z-pins penetrate both skins to create a 3D fibre structure, with the pins orientated in a tetragonal truss network as illustrated in Figure 9.15 to provide maximum resistance to shear and compression loads.

Frietas et al. (1996) report that z-pinned sandwich composites have shear and compressive strengths that are about 4 and 10 times higher than unreinforced foam, respectively. It is also claimed that z-pinned sandwich composites have higher skin-to-core bond strength, better impact damage tolerance, and are more resistant to moisture ingress than conventional honeycomb sandwich materials (Freitas et al., 1996; Palazotto et al., 1999; Vaidya et al., 1999; Vaidya, 2000). Much more work is required to assess the mechanical properties and damage resistance of z-pinned sandwich composites.

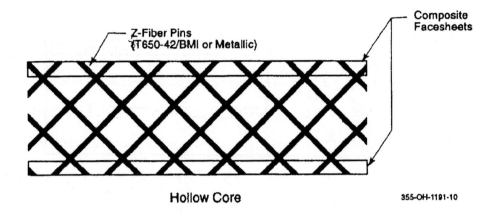

Figure 9.15 Illustration of 3D sandwich composites reinforced with z-pin.

References

A&P Technology, 1997, New Dimensions, Vol. 1, Issue 1, Spring, pp 3.

Aboudi J., 1991, Mechanics of Composite Materials – A Unified Micromechanical Approach, Elsevier, Amsterdam, The Netherlands.

Addis S., 1996, A bias yarn assembly forming device, Int. Patent WO 96/06213, 29 February.

Anahara M., Y. Yasui and H. Omori, 1991, Three-dimensional textile and method of producing the same, European Patent 0-426-878-A1, 15 May.

Anwar K.O., P. J. Callus, K. H. Leong, J. I. Curiskis and I. Herszberg, 1997, Proc. of the 11th Int. Conf. on Composite Materials, July 14-18, Gold Coast, Australia, 5:328.

Arendts F.J., K. Drechler and J. Brandt, 1993, Advanced textile structural composites – status and outlook, Proc. of the Int. Conf. on Advanced Composite Materials, Ed T. Chandra and A.K. Dhingra, 409-416.

Aymerich F., P. Priolo, R. Sanna and C.T. Sun, 2001, Fatigue behaviour of stitched composite laminates', Proc. 13th Int. Conf. Comp. Mats., 25-29 June, Beijing, Paper No. 1461.

Bannister M.K., 2001, Challenges for composites into the next millennium – a reinforcement perspective, Comp., 32A:901.

Bannister M.K., R. Braemar and P. Crothers, 1999, The mechanical performance of 3D woven sandwich structures, Comp. Struc., 47:687-690.

Bannister M.K., I. Herszberg, A. Nicolaidis, F. Coman and K.H. Leong, 1998, Composites Part A, 29A:293-300.

Bannister M.K. and A. Nicolaidis, 1998, The development of textile composite prototypes for structural applications, Proc. of the 4th Int. Conf. on Textile Composites, Kyoto, Japan, 12-14 October, pp O-36-1 – O-36-6.

Bathgate R.G., C.H. Wang and F. Pang, 1997, Effects of temperature on the creep behaviour of woven and stitched composites, Comp. Struct., 38:435.

Bauer J., 2000, Cost improvement by changing the technology, Proc. of the 1st Stade Composite Colloquium, September 7-8, Stade, Germany, 305-318.

Beckworth S.W. and C.R. Hyland, 1998, Resin transfer moulding: a decade of technology advances, SAMPE J., 34:7-19.

Bibo G.A. and P.J. Hogg, 1996, The role of reinforcement architecture on impact damage mechanisms and post-impact compression behaviour, J. Mat. Sci., 31:1115.

Bibo G.A., P.J. Hogg and M. Kemp, 1997, Mechanical characterisation of glass- and carbon-fibre-reinforced composites made with non-crimp fabrics, Comp. Sci. & Tech., 57:1221-1241.

Bibo G.A., P.J. Hogg, R. Backhouse and A. Mills, 1998, Carbon-fibre non-crimp fabric laminates for cost-effective damage-tolerant structures, Comp. Sci. & Tech., 58:129-143.

Billaut F. and O. Roussel, 1995, Impact resistance of 3-D graphite/epoxy composites, Proc. of the 11st Int. Conf. on Comp. Mat., Vol. 5, Ed. A. Pourartip and K. Street, Whistler, BC, Canada, 14-18 August 1995, Woodhead Publishing Ltd., 551-558.

Boyce J., R. Wallis, Jr. and D. Bullock, 1989, Composite Structure Reinforcement, US Patent No. 4,808,461, February 28.

Brandt J. and K. Drechsler, 1995, The potential of advanced textile structural composites for automotive and aerospace applications, Proc. of the 4th Japan Int. SAMPE Symposium, 25-28 September, 679-686.

Brandt J., K. Drechsler and F.-J. Arendts, 1996, Mechanical performance of composites based on various three-dimensional woven-fibre preforms, Comp. Sci. & Tech., 56:381-386.

Brookstein D.S., 1990, Interlocked fiber architecture: Braided and woven, Proc. of the 35th Int. SAMPE Symposium, 2-5 April, 746-756.

Brookstein D.S., 1991a, Comparison of multilayer interlocked braided composites with other 3D braided composites, Proc. of the 36th Int. SAMPE Symposium, 15-18 April, 141-150.

Brookstein D, 1991b, Evolution of fabric preforms for fiber composites, Journal of Applied Polymer Science: Applied Polymer Symposium, 47:487-500.

Brookstein D.S., 1993, Three-dimensional braids for reinforcing composites, Tech. Text. Int., May, 12-14.

Brookstein D., T. Preller and J. Brandt J, 1993, On the mechanical properties of three-dimensional multilayer interlock braided composites, Proc. of the Techtextil Symposium 93 for Technical Textiles and Textile-Reinforced Materials, Frankfurt am Main, Germany, June 7-9, Vol. 3.2, paper 3.28.

Broslus D. and S. Clarke, 1991, Textile preforming techniques for low cost structural composites, Proc. of Advanced Composite Materials: New developments and Applications Conf., Detroit, USA, September 30-October 3, 1-10.

Brown A.S., 1997, Cutting composite costs with needle and thread, Aerospace America, Nov., 24.

Brown R., 1985, Through-the-thickness braiding technology, Proc. of the 30th national SAMPE Symposium, 19-21 March, 1509-1518.

Brown R., 1988, Braiding Apparatus, US Patent 4,753,150, June 28.

Brown R.T., 1991, Design and manufacturing of 3-D braided preforms, Proc. of the 5th Textile Structural Composites Symposium, Philadelphia, Dec. 5, 1991.

Brown R. and E. Crow, 1992, Automated through-the-thickness braiding, Proc. of the 37th National SAMPE Symposium, 9-12 March, 832-841.

Bruno P., D. Keith and A. Vicario, 1986, Automatically woven three-dimensional composite structures, SAMPE Quarterly, 17:10-17

Byun J-H. and T-W. Chou, 1991, MD-Vol 29, Plastics and Plastic Composites: Material Properties, Part Performance and Process Simulation, ASME, 309.

Byun J-H, Gillespie, J.W. and Chou, T-W., 1989. Mode I delamination of a three-dimensional fabric composite, Delamination in Advanced Composites, Ed. G.M. Newaz, Technomic Publishing Co., 457-478.

Byun J.H., T.J. Whitney, G.W. Du and T.W. Chou, 1991, Analytical characterization of two-step braided composites, J. Comp. Mat., 25:1599-1618.

Cacho-Negrete C., 1982, Integral Composite Skin and Spar (ICSS) Study Program, AFWAL-TR-82-3053, Flight Dynamics Laboratory, Wright Aeronautical Laboratories, Wright-Patterson Air Force Base, Ohio.

Callus P.J., A.P. Mouritz, M.K. Bannister and K.H. Leong, 1999, Tensile properties and failure mechanisms of 3D woven GRP composites, Composites, 30A, 1277-1287.

Caneva C., S. Olivieri, C. Santulli and G. Bonifazi, 1993, Impact damage evaluation on advanced stitched composites by means of acoustic emission and image analysis, Comp. Struct., 25:121.

Caprino G., 1984, Residual strength predicted in impacted CFRP laminates, J. Comp. Mat., 18:508-518

Cartié D.D.R. and I.K. Partridge, 1999a, Delamination behaviour of Z-pinned laminates, Proc. 12th Int. Conf. Comp. Mat., 5-9 July, Paris.

Cartié D.D.R. and I.K. Partridge, 1999b, Delamination behaviour of Z-pinned laminates, Proc. 2nd ESIS TC4 Conf., Ed. J. G. Williams, 13-15 Sept., CH- Les Diablerets, Elsevier.

Chamis C.C., 1984, Simplified composite micromecahnics equations for hygral thermal and mechanical properties, SAMPE Quarterly, April, 14-23

Chapman C. and J. Whitcomb, 1995, Effect of assumed tow architecture on predicted moduli and stresses in plain weave composites, J. Comp. Mat., 29:2134-2159.

Chen L., X. M. Tao and C. L. Choy, 1999, On the microstructure of three-dimensional braided preforms, Comp. Sci. & Tech., 59:391-404.

Chiu C.-H., C.-K. Lu and T.-C. An, 1994, The influence of braiding pitch-length on flexural behaviour of 2-step braiding I-beam composites, Proc. of the 39th Int. SAMPE Symposium., 11 - 14 April, 1617-1628.

Cholakara M.T., B.Z. Jang and C.Z. Wang, 1989, Deformation and failure mechanisms in 3D composites, Proc. 34th Int. SAMPE Symp., 8-11 May, 2153-2160.

Chou S., H.-C. Chen and H.-E. Chen, 1992, Effect of weave structure on mechanical fracture behaviour of three-dimensional carbon fiber fabric reinforced epoxy resin composites, Comp. Sci. & Tech., 45:23-35.

Chou S., H-C Chen and C.-C. Wu, 1992, BMI resin composites reinforced with 3D carbon-fibre fabrics', Comp. Sci. & Tech., 43:117-128.

Chou S. and C-J. Wu, 1992, A study of the physical-properties of epoxy-resin composites reinforced with knitted glass-fiber fabrics, J. Reinforced Plastics & Comp., 11:1239-1250.

Chou T.W. and F.K. Ko, 1989, Textile Structural Composites, Volume 3 Composite Materials Series, Elsevier Science Publishers, Amsterdam, New York, U.S.A., 1989.

Christensen R.M., 1979, Mechanics of Composite Materials, John Wiley, USA

Christensen R.M., 1990, A critical evaluation for a class of micromechanics models, J. Mech. Phy. Solids, 38:379-404.

Clayton G., P. Falzon, S. Georgiadis and X.J. Liu, 1997, Towards a composite civil aircraft wing, Proc. of the 11th Int. Conf. on Composite Materials, ICCM-11, 14-18 July, I-310-319.

Cox B.N., 1999, Constitutive model for a fiber tow bridging a delaminated crack, Mech. Comp. Mat. & Struct., 6:117.

Cox B.N., 2000, Simple, conservative criteria for buckling and delamination propagation in the presence of stitching, J. Comp. Mat, 34:1136.

Cox B.N., W.C. Carter and N.A. Fleck, 1994, A Binary Model of Textile Composites-I. Formulation, Acta Metall. Mater., 42:3463-3479.

Cox B.N. and M.S. Dadkhah, 1995, The Macroscopic Elasticity of 3D Woven Composites, J. Comp. Mat., 29:785-819.

Cox B.N., M.S. Dadkhah, R.V. Inman, W.L. Morris and J. Zupon, 1992. Mechanisms of compressive failure in 3D composites, Acta Metal. et Mat., 40:3285-3298.

Cox B.N., M.S. Dadhkak, W.L. Morris and J.G. Flintoff, 1994, Failure mechanisms of 3D woven composites in tension, compression and bending, Acta Metal. et Mat., 42:3967-3984.

Cox B.N. and G. Flanagan, 1996, Handbook of Analytical Methods for Textile Composites, Rockwell Science Center.

Cox B.N., R. Massabò, D. Mumm, A. Turrettini and K. Kedward, 1997, Delamination fracture in the presence of through-thickness reinforcement, Proc. 11th Int. Conf. Comp. Mat., 14-18 July, Ed. M. L. Scott, Technomic Publishing, Lancaster, Pennsylvania, I-159 to I-177.

Crane R.M. and E.T. Camponesch, 1986, Experimental and analytical characterization of multidimensionally braided graphite/epoxy composites, Exp. Mech., Sep., 259-266.

Curiskis J., A. Durie, A. Nicolaidis and I. Herszberg, 1997, Developments in multiaxial weaving for advanced composite materials, Proc. of the 11th Int. Conf. on Composite Materials, Vol. 5, Gold Coast, Australia, Jul 14-18, 87-96.

Dadkhah M.S., W.L. Morris and B.N. Cox, 1995, Compression-compression fatigue in 3D woven composites, Acta Metal. et Mat., 43:4235-4245.

Daniel I.M. and O. Ishai, 1994, Engineering Mechanics of Composite Materials, Oxford University Press, Oxford.

de Haan, J., K. Kameo, A. Nakai, A. Fujita, J. Mayer and H. Hamada, 1997, Proc. of the 5th Japan Int. SAMPE Symposium, Oct 28-31, pp 641.

Deaton J.W., S.M. Kullerd, R.C. Madan and V.L. Chen, 1992, Test and analysis results for composite transport fuselage and wing structures, Proc. Fiber-Tex 1992, 27-29 Oct., NASA Conf Pub., 3211, 169-193.

Dexter H.B., 1992, An overview of the NASA textile composites program, Proc. of Fiber-Tex 1992, 27-29 Oct., NASA Conf. Pub., 3211, 1-31.

Dexter H.B., 1996, Innovative textile reinforced composite materials for aircraft structures, Proc. of the 28th Int. SAMPE Technical Conf., 4-7 Nov., 404-416.

Dexter H. B. and G. H. Hasko, 1996, Mechanical properties and damage tolerance of multiaxial warp-knit composites, Comp. Sci. and Tech., 56:367-380.

Dexter H.B. and J.G. Funk, 1986, Impact resistance and interlaminar fracture toughness of through-the-thickness reinforced graphite/epoxy, AIAA paper 86-CP, 700-709.

Ding Y.Q., W. Wenger and R. McIlhagger, 1993, Structural characterization and mechanical properties of 3-D woven composites, Proc. of European SAMPE Conf.

Dorey G., 1989, Damage tolerance and damaged assessment in advanced composites, in Advanced Composites, Ed. K. Partridge, Elsevier, Amsterdam, chapter 11.

Dow M.B. and H.B. Dexter, 1997. NASA Technical Publication TP-97-206234.

Dow M.B. and D.L. Smith, 1989, Damage-tolerant composite materials produced by stitching carbon fabrics, Proc. 21st Int. SAMPE Tech. Conf., 25-28 Sept., 595-605.

Dransfield K.A., 1995, Through-thickness reinforcement of carbon fibre composites by stitching, PhD Thesis, University of Sydney.

Dransfield K., C. Baillie and Y.-W. Mai, 1994, Improving the delamination resistance of CFRP by stitching - a review, Comp Sci & Tech, 50:305.

Dransfield K.A., M.G. Bader, C.A. Baillie and Y.-W. Mai, 1995, The effect of cross-stitching with an Aramid yarn on the delamination fracture toughness of CFRPs, Proc. 3rd Int. Conf. Deformation & Fracture of Composites, 27-29 Mar., 414-423.

Dransfield K.A., L.K. Jain and Y.-W. Mai, 1998, On the effects of stitching in CFRPs – I. Mode I delamination toughness, Comp. Sci. & Tech., 58:815.

Du X., F. Xue and Z. Gu, 1986, Experimental study of the effect of stitching on strength of a composite laminate, Proc. Int. Symp. Comp. Mat. & Struct., 10-13 Jun., 912-918.

Editor 1996, Total garment technology – the pursuit of WholeGarment production, Knitting Int., December/January.

Editor 1997, Flat frame machines, Knitting Int., March, 30-31.

Emehel T.C. and K.N. Shivakumar, 1997, Tow collapse model for compression strength of textile composites, J. Rein. Plastics & Comp., 16:86-101.

Epstein M. and S. Nurmi, 1991, Near net shape knitting of fiber glass and carbon for composites, Proc. of the 36th Int. SAMPE Symposium, April 15-18, 102-113.

Evans D. and J. Boyce, 1989, Transverse reinforcement methods for improved delamination resistance, 34th Int. SAMPE Symposium, May 8-11, 271-282.

Farley G.L., 1992, A mechanism responsible for reducing compression strength of through-the-thickness reinforced composite material, J. Comp. Mat., 26:1784.

Farley G., 1993, Method and apparatus for weaving a woven angle ply fabric, US Patent 5,224,519, 6 July.

Farley G.L. and Dickinson, L.C., 1992, Removal of surface loop from stitched composites can improve compression and compression-after-impact strengths, J. Rein. Plast. & Comp., 11:633.

Farley G.L., B.T. Smith and J. Maiden, 1992, Compression response of thick layer composite laminates with through-the-thickness reinforcement, J. Rein. Plast. & Comp., 11:787-810.

Fedro M.J. and K. Willden, 1991, 'Characterization and manufacture of braided composites for large commercial aircraft structures', NASA CP 3154, Second NASA Advanced Composites Technology Conf., 4-7 Nov., 387-429.

Filatovs G.J., R. L. Sadler and A. El-Shiekh, 1994, Fracture-behavior of a 3-d braid graphite-epoxy composite, J. Comp. Mat., 28:526-542.

Florentine R., 1982, Apparatus for weaving a three-dimensional article, US Patent 4,312,261, January 26.

Freitas G. and M. Dubberly, 1997, Joining aluminium materials using ultrasonic impactors, J. Metals, May, 31.

Freitas G., T. Frusco, T. Campbell, J. Harris and S. Rosenberg, 1996, Z-Fiber technology and products for enhancing composite design, Proc. of the 83rd Meeting of the AGARD SMP on "Bolted/Bonded Joints in Polymeric Composites", Sep. 2-3, Florence, Italy, pp.17-1 – 17-8.

Freitas G., C. Magee, P. Dardzinski and T. Fusco, 1994, Fiber insertion process for improved damage tolerance in aircraft laminates, J. Advanced Mat., 25:36-43.

Freitas G., C. Magee, J. Boyce and R. Bott, 1991, Service tough composite structures using z-fiber process, Proc. 9th DoD/NASA/FAA Conf. Fibrous Comp., Lake Tahoe, Nevada, Nov.

Fukuta K., Y. Nagatsuka, S. Tsuburaya, R. Miyashita, J. Sekiguti, E. Aoki and M. Sasahara, 1974, Three-dimensional fabric and method and construction for the production thereof, US Patent No. 3,834,424, Sep. 10.

Fukuta K., M. Kinbara, M. Amano, H. Tamaki, H. Ozaki, K. Nakamura, M. Furuyama, K. Mitani and T. Takei, 1995, Research and development of 3D fabric reinforced composites, Proc. of the 4th Japan Int. SAMPE Symp., 25-28 Sept., 736-741.

Fujita A., A. Nakai, A. Yokoyama and H. Hamada, 1995, Simulation system for mechanical behaviours of textile composites, Computational Mechanics'95 Theory and Applications, Proc. of the Int. Conf. on Computational Eng. Sci., Hawaii, USA, 30th July - 3rd Aug., 1995, 2311-2316.

Funk J., H.B. Dexter and S.J. Lubowinski, 1985, Experimental evaluation of stitched graphite/epoxy composites, NASA Conf. Pub. 2420, 185-205.

Furrow K.W., A.C. Loos and R.J. Cano, 1996, Environmental effects on stitched RTM textile composites, J. Rein. Plast. & Comp., 15:378.

Gause L.W. and J.W. Alper, 1987, Structural properties of braided graphite/epoxy composites, J. Comp. Tech. & Res., 9:141-150.

Gethers C. K., R. L. Saddler and V. S. Avva, 1994, Proc. of the 3rd Int. Conf. on Flow Processes in Composite Materials, July 7-9, Galway, Ireland, pp 436.

Glaessgen E.H., O.H. Griffin, C.M. Pastore and A. Birger, 1994, Modeling of textile composites, ICCE, 1, 183-184.

Glaessgen E.H., C.M. Pastore, O.H. Griffin and A. Birger, 1996, Geometrical and finite element modelling of textile composites, Composites, 27B:43-50.

Gibbon J., 1994, Knitting in the third dimension, Text. Horizon, 14:22.

Gong J.C. and Sankar B.V., 1991, Impact properties of three-dimensional braided graphite/epoxy composites, J. Comp. Mat., 25:715-731.

Graftieaux B., A. Rezai and I.K. Partridge, 2000, Effects of Z-pin reinforcement on the delamination toughness and fatigue performance of unidirectional AS4/8552 composite, Proc. ECCM9, 4-7 June, Brighton, UK.

Guénon V.A., T.-W. Chou and J.W. Gillespie, 1989, Toughness properties of a three-dimensional carbon-epoxy composite, J. Mat Sci., 24:4168-4175.

Guess T.R. and E.D. Reedy, 1985, Comparison of interlocked fabric and laminated fabric Kevlar 49/epoxy composites, J. Comp. Tech. & Res., 7:136-142.

Guess T.R. and E.D. Reedy, 1986, Additional comparisons of interlocked fabric and laminated fabric Kevlar 49/epoxy composites, J. Comp. Tech. & Res., 8:163-168.

Ha S-W., J. Mayer, J. De Haan, M. Petitmermet, E. Wintermantel, 1993, Proc. of the 6th European Conf. on Composite Materials, Sep. 20-24, Bordeaux, France, 637.

Hahn H.T. and R. Pandey, 1994, A micromechanics model for thermoelastic properties of plain weave fabric composites, J. Eng. Mat. & Tech., 116:517-523.

Hahn H.T. and S.W. Tsai, 1973, Nonlinear Elastic Behavior of Unidirectional Composite Laminae, J. Comp. Mat., 7:102-118.

Hamilton S. and N. Schinske, 1990, Multiaxial stitched preform reinforcement, Proc. of the 6th Annual ASM/ESD Advanced Composites Conf.s, 8-11 Oct., 433-434.

Harris H., N. Schinske, R. Krueger and B. Swanson, 1991, Multiaxial stitched preform reinforcements for RTM fabrication, Proc. 36th Int. SAMPE Symp., 15-18 Apr., 521-535.

Hashin Z. 1979, Analyis of properties of fibre composites with anisotropic constituents, J. Appl. Mech., 46:543-550.

He M. and B.N. Cox, 1998, Crack bridging by through-thickness reinforcement in delaminating curved structures, Comp. 29A:377.

Herrick J.W. and R. Globus, 1980, Impact resistance multidimensional composites, Proc. of the 12th National SAMPE Technical Conf., 7-9 Oct, 845-856.

Herszberg I. and M.K. Bannister, 1993, Compression and compression-after-impact properties of thin stitched carbon/epoxy composites, Proc. 5th Aust. Aero. Conf., 20-23 March.

Herszberg I., K.H. Leong and M.K. Bannister, 1995, The effect of stitching on the impact damage resistance and tolerance of uniweave carbon/epoxy laminates, Proc. 4th Int. Conf. Automated Comp., Vol. 1, 6-7 Sept., 53-60.

Herszberg I., T. Weller, K.H. Leong and M.K. Bannister, 1996, The residual tensile strength of stitched and unstitched carbon/epoxy laminates impacted under tensile load, Proc. 1st Aust. Congress on Applied Mechanics, 21-23 Feb., 309-314.

Herszberg I., A. Loh, M.K. Bannister and H.G.S.J. Thuis, 1997, Open hole fatigue of stitched and unstitched carbon/epoxy laminates. Proc. 11th Int. Conf. Comp. Mat. (ICCM-11), 14-18 July, Ed. M. L. Scott, Technomic Publishing, Lancaster, Pennsylvania, pp. V-138 to V-148.

Hill R., 1965, Theory of mechanical properties of fibre-strengthen materials – III. Self-consistent model, J. Mech. Phys. Solids, 13:189-198.

Hinrichsen J. 2000, Composite materials in the A3XX – From history to future, Proc. of the 1st Stade Composite Colloquium, pp. vii-xxxvi.

Höhfeld J., M. Drews and R. Kaldenhoff, 1994, Proc. of the 3rd Int. Conf. of Flow Processes in Composite Materials, July 7-9, 120.

Hogg P., A. Ahmadnia and F. Guild, 1993, The mechanical properties of non-crimped fabric-based composites, Comp., 24:423-432.

Holt H.B., 1992, Future composite aircraft structures may be sewn together, Auto. Eng., 90, July, 46-49.

Hranac K., 2001, Evolving preforms are the core of complex parts, High Performance Composites, September/October, 41-45.

Huang S.L., R.J. Richey and E.W. Deska, 1978, Cross reinforcement in a GR/EP laminate, American Society of Mechanical Engineers, Annual Winter Meeting, 10-15 Dec.

Huang Z.M. and S. Ramakrishna, 2000, Micromechanical modelling approaches for the stiffness and strength of knitted fabric composites: a review and comparative study, Composites, 31A:479-501.

Huang Z.M., Y. Zhang and S. Ramakrishna, 2001, Modelling of the progressive failure behavior of multilayer knitted fabric-reinforced composite laminates, Comp. Sci. & Tech., 61:2033-2046.

Huey C., 1994, Shuttle plate braiding machine, US Patent 5,301,596, 12 April.

Huysmans G., B. Gommers and I. Verpoest, 1996, Proc. of the 17th Int. SAMPE Europe Conf., Basel, Switzerland, 97.

Huysmans G., I. Verpoest and P. Van Houtte, 2001, A damage model for knitted fabric composites, Composites, 32A:1465-1475.

Hyer M.W., H.H. Lee and T.W. Knott, 1994, A simple evaluation of thermally-induced stresses in the vicinity of the stitch in a through-thickness reinforced cross-ply laminate, Center for Composite Materials and Structures, Virginia Polytechnic Institute and State University, CCMS-94-05.

Ishikawa T. and T.W. Chou, 1982a, Elastic behaviour of woven hybrid composites, J. Comp. Mat., 16:2-19.

Ishikawa T. and T.W. Chou, 1982b, Stiffness and strength behaviour of woven fabric composites, J. Mat. Sci., 17:3211-3220.

Ishikawa T. and T.W. Chou, 1983a, One-dimensional Micromechanical analysis of Woven Fabric Composites, AIAA J., 21:1714-1721.

Ishikawa T. and T.W. Chou, 1983b, In-plane thermal expansion and thermal bending coefficients of fabric composites, J. Comp. Mat., 17:92-104.

Ishikawa T. and T.W. Chou, 1983c, Nonlinear behaviour of woven fabric composites, J. Comp. Mat., 17:399-413.

Ishikawa T, M. Matsushima and Y. Hayashi, 1985, Experimental confirmation of the theory of elastic moduli of fabric composites, J. Comp. Mat., 19:443-458.

Jackson A.C., R.E. Barrie, B.M. Shah and J.G. Shulka, 1992, Advanced textile applications for primary aircraft structures, Proc. of Fiber-Tex 1992, 27-29 Oct., NASA Conf Pub., 3211, 325-352.

Jain L.K. and Y.-W. Mai., 1994a, On the effect of stitching on Mode I delamination toughness of laminated composites, Comp. Sci. & Tech., 51:331.

Jain L.K. and Y.-W. Mai, 1994b, Mode I delamination toughness of laminated composites with through-thickness reinforcement, App. Comp. Mat., 1:1.

Jain L.K. and Y.-W. Mai, 1994c, On the equivalence of stress intensity and energy approaches in bridging analysis, Fatigue Fract. Eng. Mat. Struct., 17:339.

Jain L.K. and Y.-W. Mai, 1994e, Analysis of stitched laminated ENF specimens for interlaminar Mode II fracture toughness, I. J. Fracture, 68:219.

Jain L.K. and Y.-W. Mai., 1995, Determination of Mode II delamination toughness of stitched laminated composites, Comp. Sci. & Tech., 55:241.

Jain L.K. and Y.-W. Mai, 1997, Recent work on stitching of laminated composites - theoretical analysis and experiments, Proc. 11th Int. Conf. Comp. Mat., 14-18 July, Ed. M. L. Scott, Technomic Publishing, Lancaster, Pennsylvania, pp.I-25 to I-51.

James B. and Howlett, S., 1997, Enhancement of post impact structural integrity of GFRP composite by through-thickness reinforcement, Proc. of the 2nd European Fighting Vehicle Symposium, 27-29 May, Shrivenham, UK.

Jegley D.C. and W.A. Waters, 1994, Test and analysis of a stitched RFI graphite-epoxy panel with a fuel access door, NASA Tech. Memo. 108992.

Jones R., 1975, Mechanics of Composite Materials, McGraw-Hill Book Co., New York

Judawisastra H., J. Ivens and I. Verpoest, 1998, The fatigue behaviour and damage development of 3D woven sandwich composites, Comp. Struct., 43:35-45.

Kamiya R., B. Cheeseman, P. Popper and T-W. Chou, 2000, Some recent advances in the fabrication and design of three-dimensional textile preforms: a review, Comp. Sci. & Tech., 60:33-47.

Kandero S.W. 2001. France, Russia to join in scramjet flight tests, Aviation Week & Space Technology, March 26, 60-62.

Kang T.J. and S.H. Lee, 1994, Effect of stitching on the mechanical and impact properties of woven laminate composite, J. Comp. Mat., 28:1574.

Keith W., 1999, Stitched composites damage tolerance performance, Boeing Phantom Works Report, Sept.

Khokar N., 1996, 3D fabric-forming processes: Distinguishing between 2D-weaving, 3D-weaving and an unspecified non-interlacing process, J. of the Textile Institute, 87, Part 1, 97-106.

Khondker O.A., K. H. Leong, M. K. Bannister and I. Herszberg, 2000, Proc. of the 32nd Int. SAMPE Technical Conf., November 5-9, Boston, USA.

Khondker O.A., K. H. Leong and I. Herszberg, 2001a, Effects of biaxial deformation of the knitted glass preform on the in-plane mechanical properties of the composite, Composites, 32A:1513-1523.

Khondker O.A., I. Herszberg and K. H. Leong, 2001b, An investigation of the structure-property relationship of knitted composites, J. Comp. Mat., 35:489-508.

Klopp K-U. Moll and B. Wulfhorst, 2000, Stitching process with one-sided approach of the textile for the production of reinforcing textiles for composites and other technical textiles, The 5th Int. Conf. on Textile Composites, September 18-20, Leuven, Belgium, 67-70.

Kim S.J., C.S. Lee, H. Shin and L. Tong, 2001, Anisotropic material characterisation of 3D orthogonal woven composite structures, AIAA-2001-1570.

Kim K-Y. and J. J. Curiskis, 2000, Private communication.

Kimbara M., M. Tsuzuki, A. Machii, M. Amano and K. Fukuta, 1995, 3D braider for textile composites, Proc. 4th Japan Int. SAMPE Symposium, 25-28 Sep, 694-699

Kimbara M., K. Fukuta, M. Tsuzuki, H. Takahama, I. Santo, M. Hayashida, A. Mori and A. Machii, 1991, Three-dimensional multi-axis fabric composite materials and methods and apparatus for making the same, US patent No. 5,076,330.

King J.E., R.P. Greaves and H. Low, 1996, Composite materials in aeroengine gas turbine: Performance potential vs commercial constraint?, Proc. of the Sixth European Conf. on Composite Materials, London, UK, May, 14-16.

Ko F.K., 1982, Three-dimensional fabrics for composites - an introduction to the Magnaweave structure, Proc. 4th Int. Conf. Composite Materials, Tokyo, 1609-1616.

Ko F.K., 1984, Developments of high damage tolerant, net shape composites through textile structural design, Proc. 5th Int. Conf. Composite Materials, 29 Jul. – 1 Aug., 1201-1210.

Ko F.K., 1989a, Textile Structural Composites, Composite Materials Series, Vol 3, Ed. T. W. Chou, F. Ko, Elsevier Science Publishers, 136-139.

Ko F.K., 1989b, Three-dimensional fabrics for composites, Composite Materials Series Vol. 13, Textile Structural Composites, eds. T-W. Chou, F. Ko, Elsevier Science Publishers, 129-171.

Ko F.K. and T.-W. Chou, 1989, Composite Materials Series 3 – Textile Structural Composites, Elsevier Science, Amsterdam, Holland

Ko F. K., J-N. Chu and C. T. Hua, 1991, J. Appl. Polymer Sci., 47:501.

Ko F.K. and D. Hartman, 1986, Impact behaviour of 2-D and 3-D glass/epoxy composites, SAMPE J., July/August, 26-30.

Ko F.K. and Pastore, C.M., 1985, Structure and properties of an integrated 3-D fabric for structural composites, Recent Advances in Composites in the United States and Japan, ASTM STPn 864, J.R. Vinson and M.Taya, Eds., American Society for Testing and Materials, Philadelphia, USA, 428-439.

Ko F.K., Pastore, C.M., Yang, J.M. and Chou, T.W., 1986, Structure and properties of multilayer multidirectional warp knit fabric reinforced composites, Proc. 3rd Japan-US Conf., Tokyo, 21-28.

Ko F., H. Soebroto and C. Lei, 1988, 3D net shaped composites by the 2-step braiding process, Proc. of the 33rd Int. SAMPE Symposium, 7-10 March, 912-921.

Kostar T. and T-W Chou, 1999, Braided Structures, 3D Textile Reinforcements in Composite Materials, ed. A. Miravete, Woodhead Pub. Ltd, Cambridge, 217-240.

Kregers A.F. and Y.G. Melbardis, 1978, Determination of three dimensional reinforced composites by the stiffness averaging method, Polymer Mechanics, 1:3-8.

Kruckenberg T. and R. Paton, Eds, 1998, Resin Transfer Moulding for Aerospace Structures, Kluwer Academic Publishers, Dordrecht.

Kullerd S.M. and M.B. Dow, 1992, Development of stitched/RTM composite primary structures, Proc. of Fiber-Tex 1992, 27-29 Oct., NASA Conf. Pub., 3211, 115-140.

Kuo W.-S. and T.-H. Ko, 2000, Compressive damage in 3-axis orthogonal fabric composites, Composites, 31A:1091-1105.

Laourine E., M. Schneider and B. Wulfhorst, 200, Production and analysis of 3D braided textile preforms for composites, Proc. of the 5th Int. Conf. on Textile Composites, Leuven, Belgium, 18-20 September, 55-58.

Larsson F., 1997, Damage tolerance of a stitched carbon/epoxy laminate, Comp. 28A:923.

Lee B., K.H. Leong and I. Herszberg, 2001, The effect of weaving on the tensile properties of carbon fibre tows and woven composites, J. Rein. Plastics & Comp., 20:652-670.

Lee S., A. Rudov-Clark and A.P. Mouritz, M. Bannister and I. Herszberg, (in press), Effect of weaving damage on the tensile properties of three-dimensional woven composites, Comp. Struct.

Lee C. and D. Liu, 1990, Tensile strength of stitching joint in woven glass fabrics, J. Eng. Mat. & Tech., 112:125-130.

Leong K. H., P. J. Falzon, M. K. Bannister and I. Herszberg, 1998, An investigation of the mechanical performance of weft-knit Milano-rib glass/epoxy composites, Comp. Sci. & Tech., 58:239-251.

Leong K.H., I. Herszberg and M.K. Bannister, 1995, Impact damage resistance and tolerance of stitched carbon epoxy laminates, Proc. 6th Aust. Aero. Conf., 20-23 March, 605-612.

Leong K.H., I. Herszberg, I. and M.K. Bannister, 1996, An investigation of fracture mechanisms of carbon epoxy laminates subjected to impact and compression-after-impact loading, Int. J. Crash, 1:285.

Leong K.H., B. Lee, I. Herszberg and M.K. Bannister, 2000, The effect of binder path on the tensile properties and failure of multilayer woven CFRP composites, Comp. Sci. & Tech., 60:149-156.

Leong K.H., M. Nguyen and I. Herszberg, 1999, The effects of deforming knitted glass fabrics on the basic composite mechanical properties, J. Mat. Sci., 34:2377-2387.

Leong K.H., S. Ramakrishna, G.A. Bibo and Z.M. Huang, 2000, The potential of knitting fbr engineering composites - a review, Composites, 31A:197-220.

Li W., A. El Shiekh, 1988, The effect of processes and processing parameters on 3-d braided preforms for composites, SAMPE Quarterly, 19:22-28.

Limmer L., G. Weissenbach, D. Brown, R. Mcllhagger and E. Wallace, 1996, The potential of 3-D woven composites exemplified in a composite component for a lower-leg prosthesis, Comp., 27A:271-277.

Liu C.H., J. Byan and T.-W. Chou, 1989, Mode II interlaminar fracture toughness of three-dimensional textile structural composites, Proc. of the Fourth Japan-US Conf. on Composite Materials, Technomic Publishing, Lancaster, PA, 981-990.

Liu D., 1987, Delamination in stitched and unstitched plates subjected to low-velocity impact, Proc. 2nd ASC Tech. Conf., Technomic Publications, 147-155.

Liu D., 1990, Delamination resistance in stitched and unstitched composite plates subjected to impact loading, J. Rein. Plastics Comp., 9:59.

Liu D., 1990, Photoelastic study on composite stitching, Exp. Tech., Feb., 25-27.

Liu H-Y and Y.-W. Mai, 2001, Effects of z-pin reinforcement on interlaminar mode I delamination, Proc. 13th Int. Conf. Comp. Mat., 25-29 June, Beijing, Paper No. 1319.

Lo, 1999, ITMA 99 Survey 4: Flat knitting machines, Textile Asia, August, 42.

Loud S. 1999. Lightweight FRP composite truss section decking developed', Adv. Mat. & Comp., 21:1-2.

Lubowinski S.J. and C.C. Poe, 1987, Fatigue characterization of stitched graphite/epoxy composites, Proc. Fibre-Tex 1987 Conf., NASA Conf. Pub. 3001, 253-271.

Lundblad W.E., Dixon, C. and Ohler, H.C., 1995, Ballistic resistant article comprising a three dimensional interlocking woven fabric, US Patent 5,456,974, 10 Oct.

Ma C.L., J.M. Yang and T.-W. Chou, 1986, Elastic stiffness of three-dimensional braided textile structure composites, Composite Materials: Testing and Design (7th Conf.), ASTM 893, 404-421.

Macander A.B., R.M. Crane and E.T Camponeschi, 1986, Fabrication and mechanical properties of multidimensionally (X-D) braided composite materials, Composite Materials: Testing and Design (7 Conf.), ASTM STP 893, American Society for Testing and Materials, Philadelphia, 422-443.

Maclander A.B., 1992, An X-D braided composite marine propeller, Proc. of the 10th DOD/NASA/FAA Conf. on Fibrous Composites in Structural Design, Volume II, April, pp. VII-19 to VII-34.

Massabò R. and B.N. Cox, 1999, Concepts for bridged Mode II delamination cracks, J. Mech. Phys. Solids, 47:1265.

Massabò R., D.R. Mumm and B.N. Cox, 1998 Characterizing mode II delamination cracks in stitched composites, Int. J. Fracture, 92:1.

Malkan S.R. and F.K Ko, 1989, Effect of fiber reinforcement geometry on single-shear and fracture behaviour of three-dimensionally braided glass/epoxy composite pins, J. Comp. Mat., 23:798-818.

Markus A., 1992, Resin transfer molding for advanced composite primary wing and fuselage structures, Proc. of Fiber-Tex 1992, 27-29 Oct., NASA Conf. Pub., 3211, 141-167.

McAllister L. and A. Taverna, 1975, A study of composition-construction variations in 3D carbon/carbon composites, Proc. of the 1975 Int. Conf. on Composite Materials, Geneva, Switzerland, April 7-11, 1:307-326.

McConnell R. and P. Popper, 1988, Complex shaped braided structures, US Patent 4,719,837, January 19.

Mohamed M. and A. Bilisik, 1995, Multilayer three-dimensional fabric and method for producing, US Patent 5,465,760, 14 November.

Mohamed M., Z. Zhang and L. Dickinson, 1988, Manufacture of Multilayer woven Preforms, ASME, Advanced Composites Processing Technology, MD-Vol 5, Book No. 00484, 81-89.

Morales A., 1990, Structural stitching of textile performs, Proc. 22nd Int. SAMPE Tech. Conf., 6-8 Nov., 1217-1230.

Mori T. and Tanaka K., 1973, Average stresses in matrix and average elastic energy of materials with misfitting inclusions, Acta Metalurgica, 21:571-574.

Mouritz A.P., 1995a, The flexural strength of stitched GRP laminates following underwater explosion shock loading, Proc. 10th Int. Conf. Comp. Mat., Vol. 5, 14-18 Aug., ppV-695-V-701.

Mouritz A.P., 1995b, The damage to stitched GRP laminates by underwater explosion shock loading, Comp. Sci. & Tech., 55:365.

Mouritz A.P., 1996, Flexural properties of stitched GRP laminates, Comp., 27A:525.

Mouritz A.P., 2001, Ballistic impact and explosive blast resistance of stitched composites, Comp., 32B:901.

Mouritz A. P., C. Baini and I. Herszberg, 1999, Mode I interlaminar fracture toughness properties of advanced textile fibreglass composites, Composites, 30A:859-870.

Mouritz A.P., M.K. Bannister, P.J. Falzon and K.H. Leong, 1999, Review of applications for advanced three-dimensional fibre textile composites, Comp., 30A:1445-1461.

Mouritz A.P. and B.N. Cox, 2000, A mechanistic approach to the properties of stitched laminates, Comp., 31A:1.

Mouritz A.P., J. Gallagher and A.A. Goodwin, 1996, Flexural and interlaminar shear properties of stitched GRP laminates following repeated impacts, Comp. Sci. & Tech., 57:509.

Mouritz A.P., Gellert E., Burchill P. and Challis K., 2001, Review of advanced composite structures for naval ships and submarines, Comp. Struct., 53:21-41.

Mouritz A.P., K.H. Leong and I. Herszberg, 1997, A review of the effect of stitching on the in-plane mechanical properties of fibre-reinforced polymer composites, Comp. 28A:979-991.

Mouritz A.P. and L.K. Jain, 1997, Interlaminar fracture properties of stitched fibreglass composites, Proc. 11th Int. Conf. Comp. Mat., 14-18 July, Ed. M. L. Scott, Technomic Publishing, Lancaster, Pennsylvania, V-116 to V-127.

Mouritz A.P. and L.K. Jain, 1999, Further validation of the Jain and Mai models for interlaminar fracture of stitched composites, Comp. Sci. & Tech., 59:1693.

Mouritz A.P., K.H. Leong and I. Herszberg, 1997, A review of the effect of stitching on the in-plane mechanical properties of fibre-reinforced polymer composites, Comp., 28A:979.

Mullen C. and P. Roy, 1972, Fabrication and properties description of AVCO 3D carbon/carbon cylinder materials, Proc. of the National SAMPE Symposium, Los Angeles, California, April 11-13, pp III-A-2-1 – III-A-2-8.

Müller J., A. Zulliger and M Dorn, 1994, Economic production of composite beams with 3D fabric tapes, Textile Month, Sep., 9-13.

Naik N.K., 1994, Woven Fabric Composites, Technomic, Lancaster, PA, USA

Naik N.K. and V.K. Ganesh, 1992, Prediction of on-axes elastic properties of plain weave fabric composites, Comp. Sci. & Tech., 45:135-152.

Naik N.K. and P.S. Shembekar, 1992a, Elastic behaviour of woven fabric composites: I-lamina analysis, J. Comp. Mat., 26:2197-2225.

Naik N.K. and P.S. Shembekar, 1992b, Elastic behaviour of woven fabric composites: III-laminate design, J. Comp. Mat., 26:2523-2541.

Naik R.A., P.G. Ifju and J.E. Masters, 1994, Effect of fibre architecture parameters on deformation fields and elastic moduli of 2-D braided composites, J. Comp. Mat., 28:656-681.

Ogo Y., 1987, The effect of stitching on in-plane and interlaminar properties of carbon epoxy fabric laminates, Report CCM-87-17, University of Delaware.

Owsley G.S., 2000, The effect of Z-fiber™ reinforcement on fatigue properties of stiffened panel panels, Proc. 15th Tech. Conf. American Soc. Comp., 25-27 Sept., Texas.

Palazotto A.N., L.N.B. Gummadi, U.K. Vaidya and E.J. Herup, 1999, Low velocity impact damage characteristics of Z-fiber reinforced sandwich panels – an experimental study, Comp. Struct., 43:275.

Palmer R.J., M.B. Dow and D.L. Smith, 1991, Development of stitching reinforcement for transport wing panels, Proc. of the 1st NASA Advanced Composites Technical Conf., Part 2, 621-646.

Parnas R. S., 2000, Liquid Composite Moulding, Hanser Publishers, Munich.

Pang F., C.H. Wang and R.G. Bathgate, 1997, Creep response of woven fibre composites and the effect of stitching, Comp. Sci. & Tech., 57:91.

Pang F., C.H. Wang and R.G. Bathgate, 1998, Modelling of the creep behaviour of stitching and un-stitched woven composites, Proc. Mats. 98, Ed. M. Ferry, 735-740.

Partridge I.K. and D.D.R. Cartié, 2001, Increasing delamination resistance of composites by z-pinning, Proc. ACUN-3, 6-9 Feb., Sydney, 169-172.

Pastore C.M. and Y.A. Gowayed, 1994, A self-consistent fabric geometry model: modification and application of a fabric geometry model to predict the elastic properties of textile composites, J. Comp. Tech. & Res., 16:32-36.

Pelstring R.M. and R.C. Madan, 1989, Stitching to improve damage tolerance of composites, Proc. 34th Int. SAMPE Symp., 8-11 May, 1519-1528.

Phillips E., 2000, Composite wing box tested to failure, Aviation Week & Space Technology, June 19, 37.

Portanova M.A., C.C. Poe and J.D. Whitcomb, 1992, Open hole and postimpact compressive fatigue of stitched and unstitched carbon-epoxy composites, In Composite Materials: Testing and Design (Tenth Volume), ASTM STP 1120, Ed. G.C. Glenn, American Society for Testing and Materials, Philadelphia, 37-53.

Popper P. 1991. Braiding, Int. Encyclopaedia of Composites, Vol 1, Ed. S.M. Lee, VCH Publishers, 130-147.

Popper P. and R. McConnell, 1987, A new 3D braid for integrated parts manufacture and improved delamination resistance - The 2-step process, Proc. of the 32nd Int. SAMPE Symposium, 6-9 April, 92-103.

Potter K., 1997, Resin Transfer Moulding, Chapman & Hall, London.

Preller T., J. Brandt and K. Drechsler, 1990, 'Manufacturing and mechanical properties of new textile fiber preforms for structural sandwich applications', Proc. of Textile Composites in Building Construction, Part 3, 307-317.

Räckers B., 1998, Resin Transfer Moulding for Aerospace Structures, Chapter 1: Introduction to resin transfer moulding, Eds. T. Kruckenberg, R. Paton, Kluwer Academic Publishers, Dordrecht.

Räckers B., C. Howe and T. Kruckenberg, 1998, Resin Transfer Moulding for Aerospace Structures, Chapter 13: Quality and process control, Eds. T. Kruckenberg, R. Paton, Kluwer Academic Publishers, Dordrecht.

Ramakrishna S., 1997a, Analysis and modelling of plain knitted fabric reinforced composites, J. Comp. Mat., 31:52-70.

Ramakrishna S., 1997b, Characterization and modelling of the tensile properties of plain weft-knit fabric-reinforced composites, Comp. Sci. & Tech., 57:1-22.

Ramakrishna S., N. K. Cuong and H. Hamada, 1997, Tensile properties of plain weft knitted glass fiber fabric reinforced epoxy composites, J. Rein. Plastics & Comp., 16:946-966.

Ramakrishna S. and D. Hull, 1993, Energy-absorption capability of epoxy composite tubes with knitted carbon-fiber fabric reinforcement, Comp. Sci. & Tech., 49:349-356.

Ramakrishna S. and D. Hull, 1994a, Tensile behaviour of knitted carbon-fiber-fabric/epoxy laminates: I. experimental, Comp. Sci.. & Tech., 50:249-258.

Ramakrishna S. and D. Hull, 1994b, Tensile behaviour of knitted carbon-fibre-fabric/epoxy laminates: II. prediction of tensile properties,. Comp. Sci. & Tech., 50:259-268.

Reeder J.R., 1995, Stitching vs a toughened matrix: compression strength effects, J. Comp. Mat., 29:2464.

Reider J. 1996, Knitting machines for technical textiles, Int. Textile Bulletin – Nonwovens – Industrial Textiles, 1:60-63.

Reedy E.D. and T.R. Guess, 1986, Additional comparisons of interlocked fabric and laminated fabric Kevlar 49/epoxy composites, J. Comp. Tech. Res., 8:163-168.

Roberts R. and W. Douglas, 1995, 3 dimensional braiding apparatus, US Patent 5,337,647, 16 August.

Robinson F. and S. Ashton, 1994, Knitting in the third dimension – A look into the future possibilities for knitted fabric, Textile Horizons, December, 12-15.

Rolincik P., 1987, Autoweave™ - A unique automated 3D weaving technology, SAMPE Journal, September/October, 40-47.

Rossi G.T., 1989, Evaluation of 3-D reinforcements in commingled, thermoplastic structural elements, Proc. of the American Helicopter Society, 22-24 May. 509-515.

Ruan X.P. and T.-W. Chou, 1996, Experimental and theoretical studies of the elastic behavior of knitted-fabric composites, Comp. Sci. & Tech., 56:1391-1403.

Rudd C.D., M.J. Owen and V. Middleton, 1990, Mechanical properties of weft knit glass fibre/polyester laminates, Comp. Sci. & Tech., 39:261-277.

Rugg K.L., B.N. Cox, K.E. Ward and G.O. Sherrick, 1998, Damage mechanisms for angled through-thickness rod reinforcement in carbon-epoxy laminates, Comp., 29A:1603.

Ruzand J.-M. and G. Guenot, 1994, Mulitaxial three-dimensional fabric and process for its manufacture, Int. Patent WO 94/20658, 15 September.

Sawyer J.W., 1985, Effect of stitching on the strength of bonded composite single lap joints. AIAA J., 23:1744-1748.

Schneider M., B. Wulfhorst and A. Pickett, 1998, Micromodelling of yarn architecture in 3D-braids and transfer to macromodelling of composites, Proc. of the 4th Int. Conf. on Textile Composites, Kyoto, Japan, 12-14 Oct., pp O-5-1 – O-5-8.

Shah Khan M.Z. and A.P. Mouritz, 1996, Fatigue behaviour of stitched GRP laminates, Comp. Sci. & Tech., 56:695.

Shah Khan M.Z. and A.P. Mouritz, 1997, Loading rate dependence of the fracture toughness and fatigue life of stitched GRP composites. Proc. 9th Int. Conf. on Fracture (ICF9), 1-5 April, 809-818.

Sharma S.K. and B.V. Sankar, 1994, Effect of stitching on impact and interlaminar properties of graphite/epoxy laminate, Proc. 9th ASC Annual Tech. Conf., 20-22 Sept., 700-708.

Sheffer E. and T. Dias, 1988, Knitting novel 3-D solid structures with multiple needle bars, Proc. of the UMIST Textile Conf. – Textile Engineered for Performance, Manchester, UK, April, 20-22.

Shembekar P.S. and N.K. Naik, 1992, Elastic behaviour of woven fabric composites: II-Laminate analysis, J. Comp. Mat., 26:2226-2246.

Shu D. and Y.-W. Mai, 1993a, Delamination buckling with bridging, Comp. Sci. & Tech., 47:25.

Shu D. and Y.-W. Mai, 1993b, Effect of stitching on interlaminar delamination extension in composite laminates, Comp. Sci. & Tech., 49:165.

Smith B.A., P. Proctor and P. Sparaco, 1994, Airframer's pursue lower aircraft costs, Aviation Week & Space Tech., Sep 5, 57-58.

Soebroto H., C. Pastore, F. Ko, 1990, Engineering design of braided structural fibreglass composite, Proc. 6th Annual ASM/ESD Advanced Composites Conf., ASM Int., 8-11 October, 435-40.

Steeves C. and N. Fleck, 1999a, Z-pinned composite laminates: Knockdown in compressive strength, Proc. DFC5, 18-19 March, London, 60-68.

Steeves C. and N. Fleck, 1999b, Proc. 12th Int. Conf. Comp. Mat., 5-9 July, Paris.

Stewart R., 2001, Composites Technology, November/December, 33.

Stoll GmbH, 1999, Flat knitting machinery for technical textiles, Technical Textiles Int., November, 20-24.

Stover E., W. Mark, I. Marfowitz and W. Mueller, 1971, Preparation of an Omniweave-reinforced carbon-carbon cylinder as a candidate for evaluation in the Advanced Heat Shield screening program, AFML-TR-70-283, March.

Su K.B., 1989, Delamination resistance of stitched thermoplastic matrix composite laminates, Advances in Thermoplastic Matrix Composite Materials, ASTM STP 1044, Ed. G.M. Newaz, American Society for Testing and Materials, Philadelphia, 279-300.

Suarez J. and S. Dastin, 1992, Comparison of resin film infusion, resin transfer moulding and consolidation of textile preforms for primary aircraft structure, Proc. Fiber-Tex 1992, 27-29 Oct., NASA Conf. Pub., 3211, 353-386.

Susuki I. and T. Takatoya, 1997, Impact damage properties of 3-D carbon/bismaleimide composites, Proc. of the 5th Japan Int. SAMPE Symposium, Tokyo, 28-31 Oct., 691-696.

Tada Y. and T. Ishikawa, 1989, Experimental evaluation of the effects of stitching on CFRP laminate specimens with various shapes and loadings, Key Eng. Mat., 37:305-316.

Tan P., 1999, Micromechanics models for mechanical properties and failure strengths of 2D and 3D woven composites, PhD Thesis, University of Sydney, Australia

Tan P., L. Tong and G.P. Steven, 1997a, Modelling for predicting the mechanical properties of textile composites - A review, Composites, 28A:903-922.

Tan P., L. Tong and G.P. Steven, 1997b, A three-dimensional modelling technique for predicting the linear elastic properties of open-packing woven fabric unit cell, Comp. Struct., 38:261-271.

Tan P., L. Tong and G.P. Steven, 1998, Modelling approaches for 3D orthogonal woven composites, J. Rein. Plastics & Comp., 17:545-577.

Tan P., L. Tong and G.P. Steven, 1999a, Models for predicting thermomechanical properties of 3D orthogonal woven composites, J. Rein. Plastics & Comp., 18:151.

Tan P., L. Tong and G.P. Steven, 1999b, Micro-mechanics models for mechanical and thermo-mechanical properties of 3D angle interlock woven composites, Comp., 30A:637-648.

Tan P., L. Tong, G.P. Steven and T. Ishikawa. 2000a. Behaviour of 3D orthogonal woven CFRP composites. I: Experimental investigation, Composites, 31A:259-271.

Tan P., L. Tong and G.P. Steven, 2000b, Behaviour of 3D orthogonal woven CFRP composites. II: Theoretical and FEA modelling, Composites, 31A:273-281.

Tan P., L. Tong and G.P. Steven, 2001, Mechanical behaviour for 3-D orthogonal woven E-glass/epoxy composites, J. Rein. Plastics & Comp., 20:274-303.

Tanzawa Y., N. Watanabe and T. Ishikawa, 1997, Interlaminar delamination toughness and strength of 3-D orthogonal interlocked fabric composite, Proc. of the Eleventh Int. Conf. on Composite Materials, Ed. M.L. Scott, 14-18 July, Gold Coast, Australia, V-47 to V-57.

Tarnopol'skii Y.M., V.A. Polyakor and L.G. Zhigun, 1973, Composite materials reinforced with a system of three mutually orthogonal fibres, I: Calculation of elastic characteristics, Polymer Mechanics, 5:853-860.

Thaxton C., R. Reid and A. El-Shiekh, 1991, Advances in 3-dimensional braiding, Proc. of the 5th Conf. on Advanced Engineering Fibres and Textile Structures for Composites"", NASA Conf. Publication 3176, 43-66.

Thompson D.M. and O.H. Griffin, 1990, 2-D to 3-D global/local finite element analysis of cross-ply composite laminates, J. Rein. Plastics & Comp. 9:492-502.

Thompson D.M. and O.H. Griffin, 1992, Verification of a 2-D to 3-D global/local finite element method for symmetric laminates, J. Rein. Plastics & Comp., 11:910-931.

Thuis H.G.S.J. and E. Bron, 1996, The effect of stitching density and laminate lay-up on the mechanical properties of stitched carbon fabrics, National Lucht-en Ruimtevaartlaboratorium, Report NLR CR 96126L.

Tong L. and L.K. Jain, 1995, Analysis of adhesive bonded composite lap joints with transverse stitching, App. Comp. Mat., 2:343.

Tong L, J.K. Jain, K.H. Leong, D. Kelly and I. Hertzberg, 1998a, Failure of transversely stitched RTM lap joints, Comp. Sci. & Tech., 58:221-227.

Tong L, K. Cleaves, T. Kruckenberg and G.P. Steven, 1998b, Effect of transverse stitching on the fracture loads of RTM single lap joints, Key Eng. Mat., 137:195.

Tong L., P. Tan and G.P. Steven, 2002, Effect of yarn waviness on strength of 3D orthogonal woven CFRP composite materials, J. Rein. Plastics & Comp., 21:153.

Tsai S.W. and E.M. Wu, 1971, A general theory of strength for anisotropic materials, J. Comp. Mat., 5:58-80.

Turrettini A., 1996, An investigation of the mode I and II stitch bridging laws in stitched polymer composites, Masters Thesis, Department of Mechanical & Environmental Engineering, University of California, Santa Barbara.

Vaidya U.K., M.V. Kamath, M.V. Hosur, M.V., H. Mahfuz and S. Jeelani, 1999, Low-velocity impact response of cross-ply laminated sandwich composites with hollow and foam-filled z-pin reinforced core, J. Comp. Tech. & Res., 21:84.

Vaidya U.K., A.N. Palazotto and L.N.B. Gummadi, 2000, Low velocity impact and compression-after-impact response of z-pin reinforced core sandwich composites, Trans. ASME, 122:434.

Van Vuure A.W., J. Ivens and I. Verpoest, 1994, Sandwich panels produced from sandwich-fabric preforms, Proc. of the Int. Symposium on Advanced Materials for Lightweight Structures, ESTEC, 609-612.

Vandermey N.E., D.H. Morris and J.E. Masters, 1991, Damage development under compression-compression fatigue loading in a stitched uniwoven graphite/epoxy composite material, NASA Report PB91-236026.

Vandeurzen P., J. Ivens and I. Verpoest, 1996a, A three-dimensional micromechanical analysis of woven fabric composites: I. Geometric analysis, Comp. Sci. & Tech., 56:1303-1315.

Vandeurzen P., J. Ivens and I. Verpoest, 1996b, A three-dimensional micromechanical analysis of woven fabric composites: II. Elastic analysis, Comp. Sci. & Tech., 56:1317-1327.

Vandeurzen P., J. Ivens and I. Verpoest, 1998, Micro-stress analysis of woven fabric composites by multilevel decomposition, J. Comp. Mat., 32:623-651.

Verpoest I., J. Dendauw, 1992, Proc. of the 37th Int. SAMPE Symposium, March 9-12, 369.

Verpoest I., B. Gommers, G. Huymans, J. Ivens, Y. Luo, S. Pandita and D Phillips, 1997, The potential of knitted fabrics as a reinforcements for composites, Proc. of the 11th Int. Conf. on Composite Materials, 14-18 July, 1-108 to 1-133.

Verpoest I., J. Ivens. A. van Vuure, B. Gommers, P. Vandeurzen, V. Efstratiou and D. Phillips, 1995, New developments in advanced textiles for composites, Proc. of the 4th Japan Int. SAMPE Symposium, September 25-28, 644-654.

Voss S., A. Fahmy and H. West, 1993, Impact tolerance of laminated and 3-dimensionally reinforced graphite-epoxy panels, Proc. of the Int. Conf. on Advanced Composite Materials, Ed T. Chandra and A.K. Dhingra, 591-596.

Vuure Van, F. Ko and R. Balonis, 1999, Textile preforming for complex shape structural composites, Proc. of the 44th Int. SAMPE Symposium, May 23-27, 293-302.

Wadsworth M., 1998, Resin Transfer Moulding for Aerospace Structures, Chapter 9: Tooling fundamentals for resin transfer moulding, Eds. T. Kruckenberg, R. Paton, Kluwer Academic Publishers, Dordrecht.

Wang Y., Y. Gowayed, X. Kong, J. Li and D. Zhao, 1995a, Properties and analysis of composites reinforced with e-glass weft-knitted fabrics, J. Comp. Tech. & Res., 17:283-288.

Wang Y., J. Li and P. B. Do, 1995b, Properties of composite laminates reinforced with e-glass multiaxial non-crimp fabrics, J. Comp. Mat., 29:2317-2333.

Watt A., A.A. Goodwin and A.P. Mouritz, 1998, Thermal degradation of the interlaminar fracture properties of stitched glass/vinyl ester composites, J. Mat. Sci., 33:2629.

Wenning L., C. Guangming, Q. Xin, 1993, Proc. of the 38th Int. SAMPE Symposium, May 10-13, 1845.

Whitcomb J.D., 1989, Three-dimensional Stress Analysis of Plain Weave Composites, Paper presented at the 3rd Symposium on Composite Material: Fatique and Fracture, Orlando, Florida, USA.

Whitcomb J.D., 1991, Iterative global/local finite element analysis, Comp. & Struct., 40:1027-1032.

Whitcomb J.D and K. Woo, 1993a, Application of iterative global/local finite-element analysis. Part 1: Linear analysis, Communications Num. Meth. in Eng., 9:745-756.

Whitcomb J. and Woo, K., 1993b, Application of iterative global/local finite-element analysis. Part 2: Geometrically non-linear analysis, Commun. Num. Meth. Eng., 9:757-766.

Whitney T.J. and T.-W. Chou, 1989, Modelling of 3-D angle-interlock textile structural composites, J. Comp. Mat., 23:891-911.

Whiteside J.B., R.J. Delasi and R.L. Schulte, 1985, Measurement of preferential moisture ingress in composite wing/spar joints, Comp. Sci. & Tech., 24:123-145.

Wittig 2000, Roboted 3-dimensional stitching technology, Proc. of the 1st Stade Composite Colloquium, September 7-8, Stade, Germany, 23-38.

Wong R., 1992, Sandwich construction in the Starship, Proc. of the 37th Int. SAMPE Symposium, 9-12 March, 186-197.

Woo K. and J. Whitcomb, 1994, Global/local finite element analysis for textile composites, J. Comp. Mat., 28:1305-1321.

Wu E. and J. Liau, 1994, Impact of unstitched and stitched laminates by line loading, J. Comp. Mat., 28:1640.

Wu E. and J. Wang, 1995, Behaviour of stitched laminates under in-plane tensile and transverse impact loading, J. Comp. Mat., 29:2254.

Wu W-L., M. Kotaki, A. Fujita, H. Hamada, M. Inoda and Z-I. Maekawa, 1993, Mechanical-properties of warp-knitted, fabric-reinforced composites, J. Reinforced Plastics & Comp., 12:1096-1110.

Wulfhorst B., E. de Weldige, R. Kaldenhoff and K.-U. Moll, 1995, New developments and applications of textile reinforcements for composite materials, Proc. of the 4th Japan Int. SAMPE Symposium, 25-28 Sept., 673-678.

Xu J., B.N. Cox, M.A. Mcglockton and W.C. Carter, 1995, A Binary Model of Textile Composites-II. The Elastic Regime, Acta Metall.Mater. 43:3511- 3524.

Yamamoto T., S. Nishiyama and M. Shinya, 1995, Study on weaving method for three-dimensional textile structural composites, Proc. of the 4th Japan Int. SAMPE Symp. Sep 25-28, 655-660.

Yang J.M., C.L. Ma and T.-W. Chou, 1986, Fibre inclination model of three-dimensional textile structural composites, J. Comp. Mat., 20:472-483.

Yau S.-S., T.-W. Chou and F.K. Ko, 1986, Flexural and axial compressive failures of three-dimensionally braided composite I-beams, Comp., 17:227-232.

Zienkiewicz O.C. and R.L. Taylor, 1989, The Finite Element Method, McGraw-Hill Book Company, England, UK.

Subject Index

A

Angle interlock, 90
Applications
- braided composites 6, 10-11, 35-36
- distance fabric composites 136
- knitted composites 11
- stitched composites 11-12, 163-164
- woven composites 7-10, 107-108, 128, 136
Average strains, 66
Average stress, 66-67

B

Ballistic impact resistance
- woven composites 132
- stitched composites 199-200
Bearing performance, 154
Binary model, 99
Brick element, 97
Bridging model, 71,76,77,162
Braided composites, 100,137
- applications 6, 10-11, 35-36
- compression properties 138-144
- damage 143, 144
- fatigue 145
- flexure 138-143
- four-step braiding 25-29
- impact damage resistance 143, 144
- interlaminar fracture properties 143, 144
- interlaminar shear properties 141, 142
- manufacture 22, 23, 25-32, 46, 56-57
- microstructure 137
- modelling 145, 146
- multilayer interlock braiding 31-32
- tension properties 138-144
- two-step braiding 29-30

Braid angle, 139, 140
Braid pattern, 138
Braided preforms, 137
Braiding process, 22-32, 140-142

C

Characterisation, 63
Combi-cell, 85
Commingled composites 47
Compressive properties
- braided composites 138-144
- distance fabric composites 136
- knitted composites 154-156, 160
- stitched composites 170-176
- woven composites 123-126
- z-pinned composites 210-211
Consolidated properties, 137
Constituent phase, 63
Crack branching, 158
Cross-over model, 103
Crowding, 142

D

Damage tolerance, 144
Damage resistance, 144
Delamination (see also *interlaminar fracture properties*)
- distance fabric composites 136
- edge delamination 215
Distance fabric composites 22, 37, 133-134, 136

E

Edge condition, 138
Effective medium element, 99
Effective properties
- effective compliance, 68
- effective elastic stiffness, 68